GEOMETRY
PRACTICE WORKBOOK

GRADES 7 - 12

AMERICAN MATH ACADEMY

By H. TONG, M.Ed.

Math Instructor & Olympiad Coach

www.americanmathacademy.com

GEOMETRY

PRACTICE WORKBOOK

GRADES 7 - 12

AMERICAN MATH ACADEMY

Writer: H.Tong

Printed in the United States of America.

ISBN: 9798869116550

For questions, suggestions, or comments, please email: americanmathacademy@gmail.com

TABLE OF CONTENTS

About the Author

Mr. Tong teaches at various private and public schools in both New York and New Jersey. In conjunction with his teaching, Mr. Tong developed his own private tutoring company. His company developed a unique way of ensuring his students' success on the math section of the SAT. His students, over the years, have been able to apply the knowledge and skills they learned during their tutoring sessions in college and beyond. Mr. Tong's academic accolades make him the best candidate to teach SAT Math. He received his master's degree in Math Education. He has won several national and state championships in various math competitions and has taken his team to victory in the Olympiads. He has trained students for Math Counts, American Math Competition (AMC), Harvard MIT Math Tournament, Princeton Math Contest, the National Math League, and many other events. His teaching style ensures his students' success. He personally invests energy and time into his students and identifies His dedication to his students is evident through his students' achievements.

Acknowledgements

I would like to take the time to acknowledge the help and support of my beloved wife, my colleagues, and my students; their feedback on this book was invaluable. Without their help, this book would not be the same. I dedicate this book to my precious daughters Vera and Nora, who inspired me to take on this project.

SAMPLE TEST

1. What is the value of x in the diagram below?

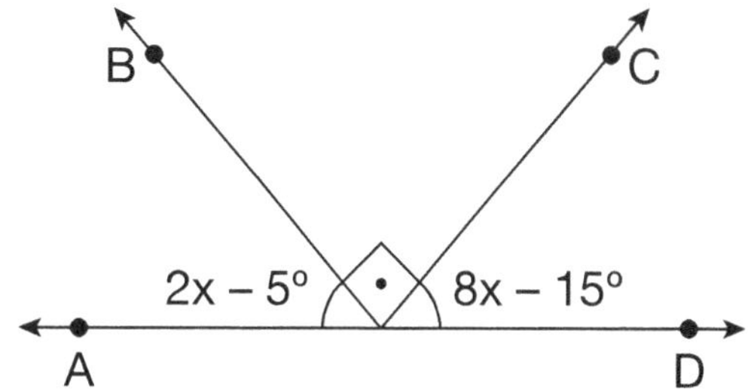

A) 5°

B) 7°

C) 9°

D) 11°

2. What is the value of x in the diagram below?

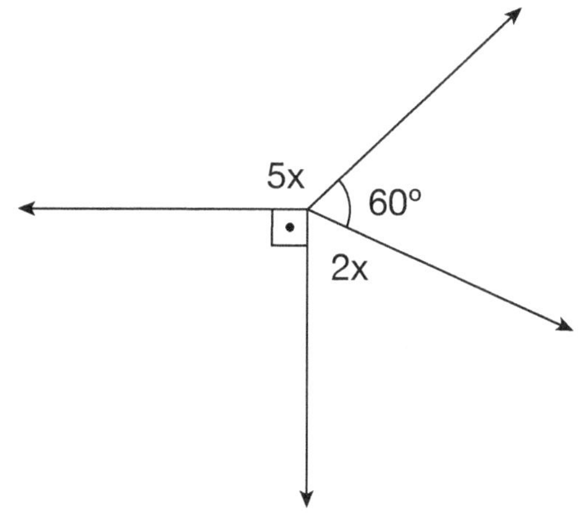

A) 20°

B) 30°

C) 40°

D) 50°

3. If C is between A and B and AB = 30,
AC = 3x + 5   CB = 2x + 10,
Find the value of x?

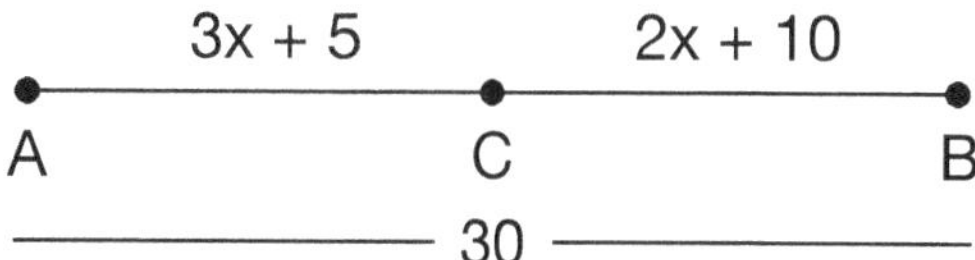

A) 1

B) 2

C) 3

D) 4

4. EF is the angle bisector of m $\angle$ DEG, if m $\angle$ DEF = 3x – 10°, and m $\angle$ FEG = 2x + 15°, then find x.

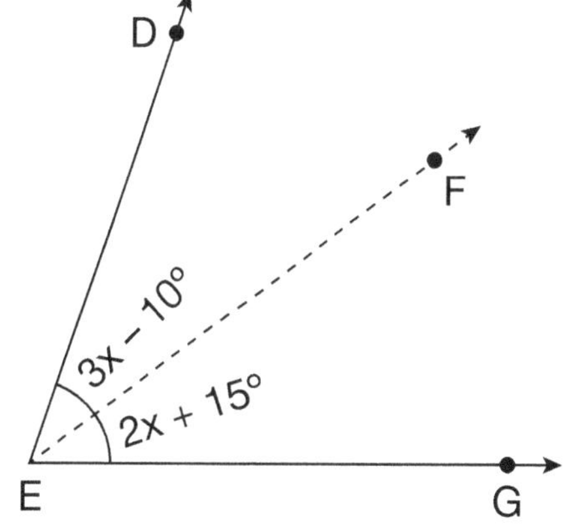

A) 10

B) 15

C) 20

D) 25

5. Find the value of x in the triangle below.

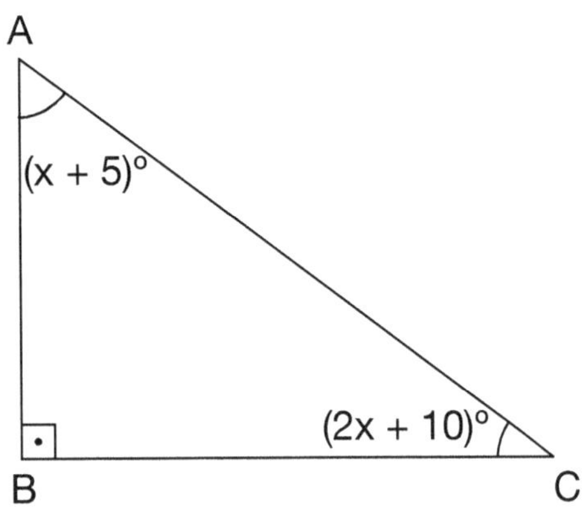

A) 20°

B) 25°

C) 30°

D) 35°

6. Find the value of x in the figure below.

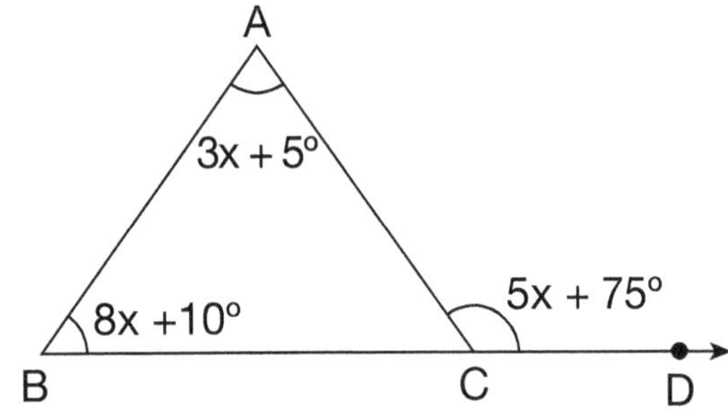

A) 3°

B) 4°

C) 7°

D) 10°

7. Which of following is the sides of $\triangle ABC$ from shortest to longest.

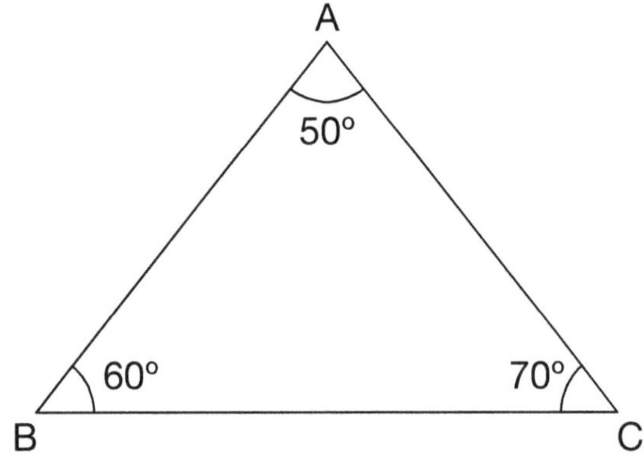

A) $\overline{AB} \angle \overline{BC} \angle \overline{AC}$

B) $\overline{BC} \angle \overline{AB} \angle \overline{AC}$

C) $\overline{AC} \angle \overline{BC} \angle \overline{AB}$

D) $\overline{BC} \angle \overline{AC} \angle \overline{AB}$

8. Find the value of x.

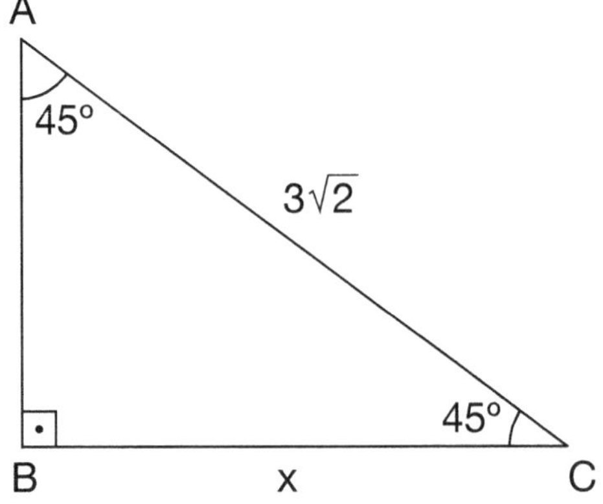

A) 1

B) 3

C) $\sqrt{3}$

D) 5

9. The midpoint between points A(6, 8) and B(a, b) is (10, 14).

What are the coordinates of point B?

A) (20, 14)

B) (14, 20)

C) (7, 10)

D) (10, 7)

10. What is the slope of the line that passes through (4, 5) (7, 11)?

A) 1

B) 2

C) 3

D) 4

11. Find the area of the following triangle.

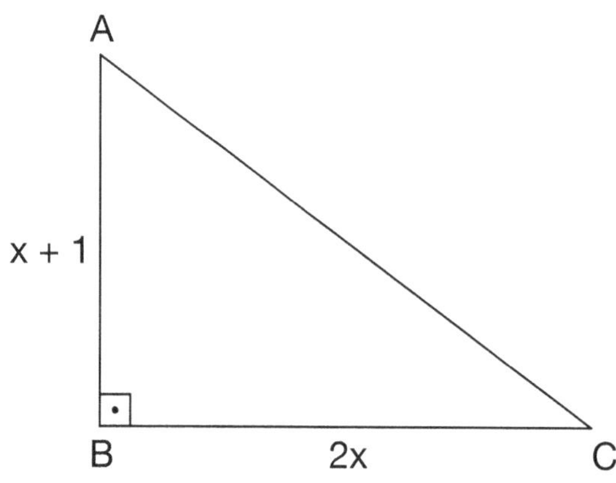

A) x^2

B) $x^2 - x$

C) $x^2 + x$

D) $2x^2 + 1$

12. What is the coordinate of the vertices of (4, –5) when the vertices reflects across the y-axis.

A) (4, 5)

B) (–4, 5)

C) (4, –5)

D) (–4, –5)

13. 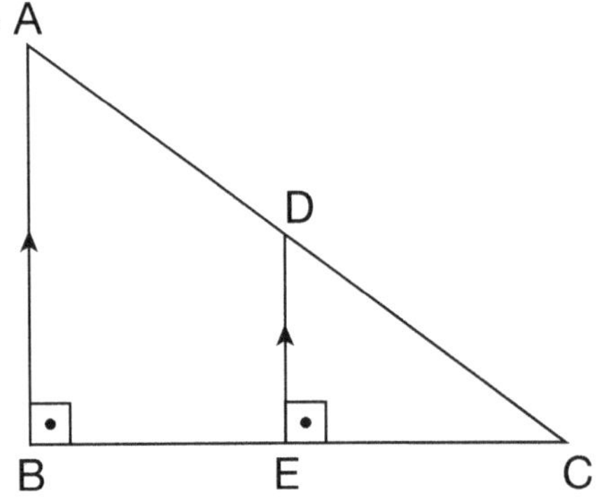

If AB // DE and |AB| = 8, |BC| = 6 and |DE| = 4 then find |DC| + |EC|?

A) 5

B) 6

C) 7

D) 8

14. In the following figure, what is the measure of ∠ ACD?

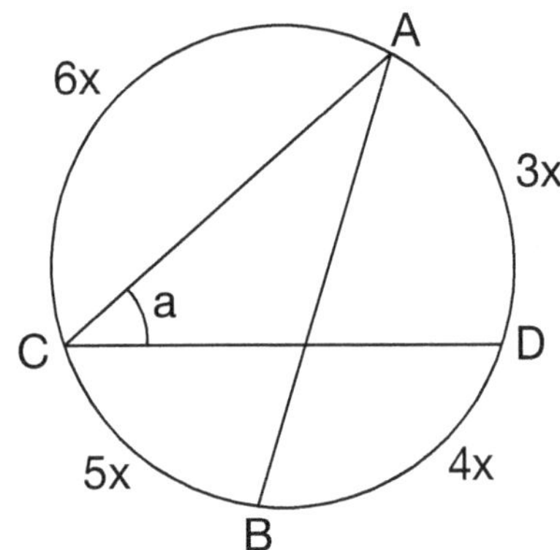

A) 15°

B) 25°

C) 30°

D) 45°

15. 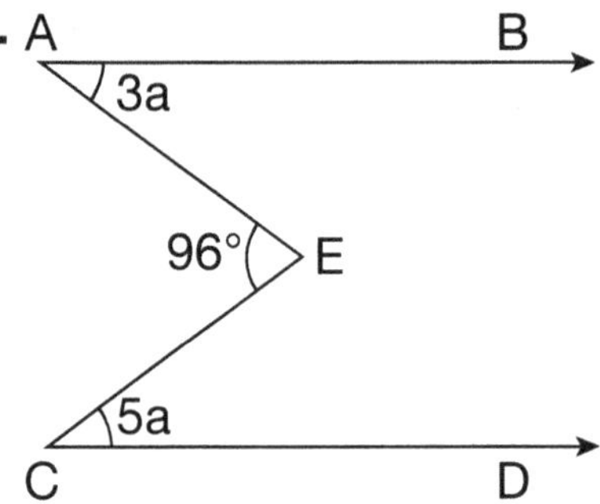

In the figure above, AB is parallel to CD. What is the angle of ∠ BAE?

A) 20°

B) 30°

C) 36°

D) 48°

16. 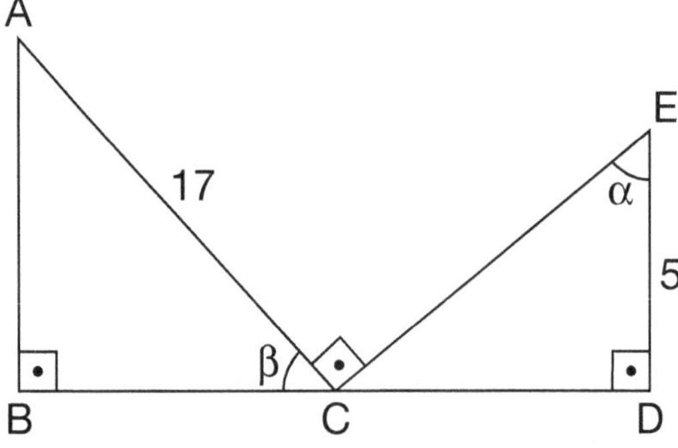

If AB ⊥ BC, ED ⊥ CD, AC ⊥ CE

|BD| = 27, |AB| = 8, |AC| = 17, |ED| = 5

From the above triangles, find Sin α • Cot β?

A) $\frac{29}{45}$

B) $\frac{45}{29}$

C) $\frac{26}{45}$

D) $\frac{45}{26}$

SAMPLE TEST

17.

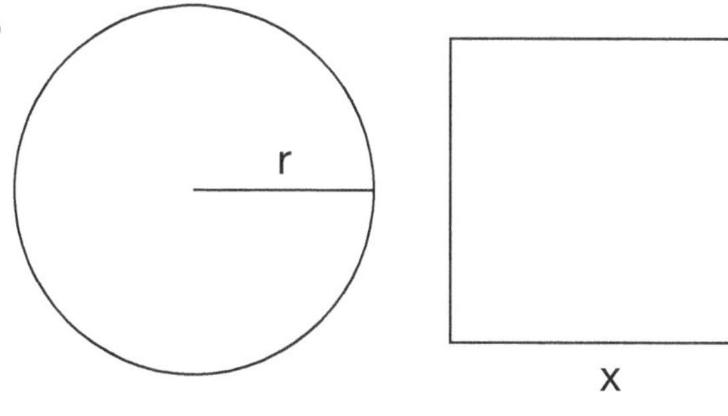

The area of the circle with radius r is equal to the area of the square with a side length of x. What is the ratio of the square's perimeter to the circle's circumference?

A) $\frac{2\sqrt{\pi}}{\pi}$

B) $\frac{\pi}{3}$

C) $\frac{3}{\pi}$

D) $\frac{2}{\pi}$

18. What is the area of an equilateral triangle with a side of 4?

A) $\sqrt{3}$

B) $2\sqrt{3}$

C) $4\sqrt{3}$

D) $6\sqrt{3}$

19.

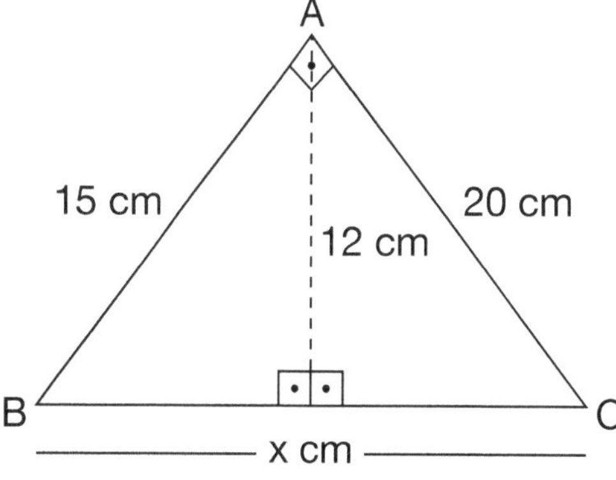

What is the value of x?

A) 15 cm

B) 20 cm

C) 25 cm

D) 30 cm

20.

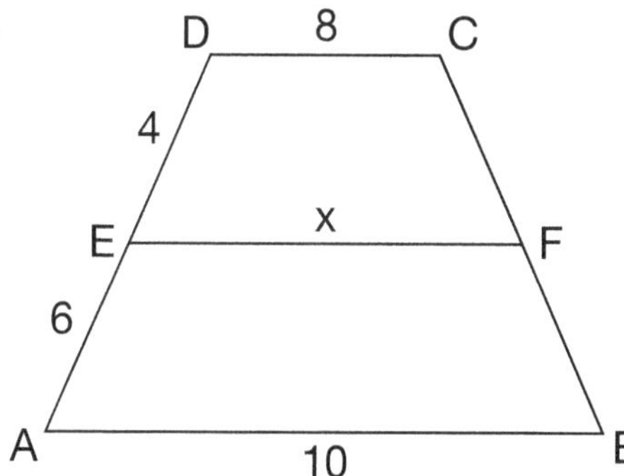

In the trapezoid ABCD above, DC = 8, EF = x, and AB = 10.

What is the value of x?

A) 2.2

B) 4.4

C) 8.8

D) 9.8

21.

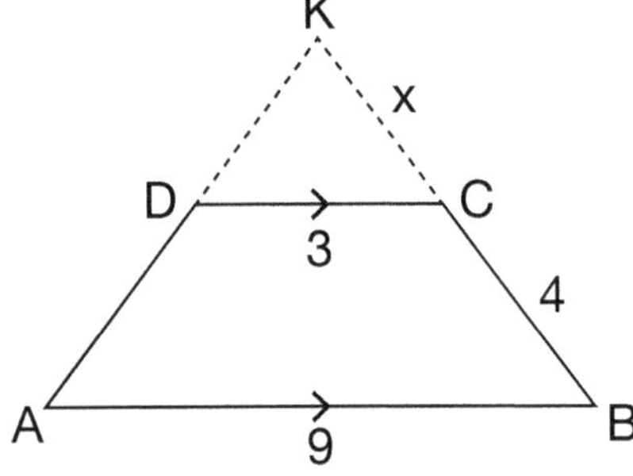

If DC // AB in the figure above, what is the value of x?

A) 2

B) 4

C) 6

D) 8

22.

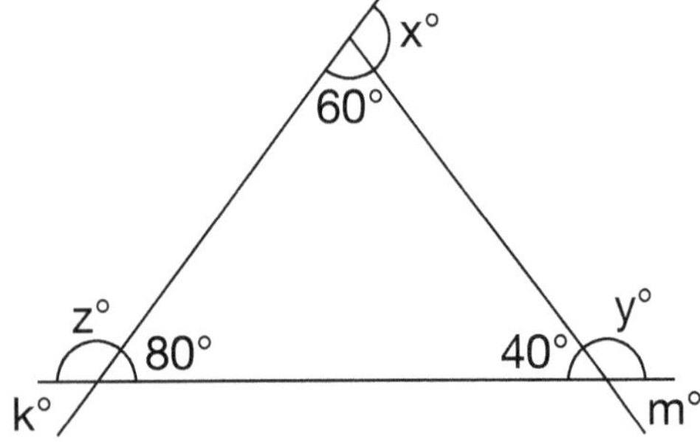

In the figure above, wich of the following is the greatest?

A) x°

B) y°

C) z°

D) k°

23.

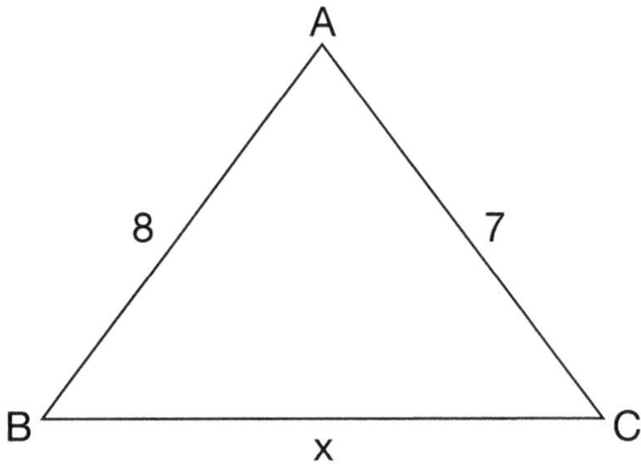

Which of following can be x?

A) 1

B) 15

C) 16

D) 12

24.

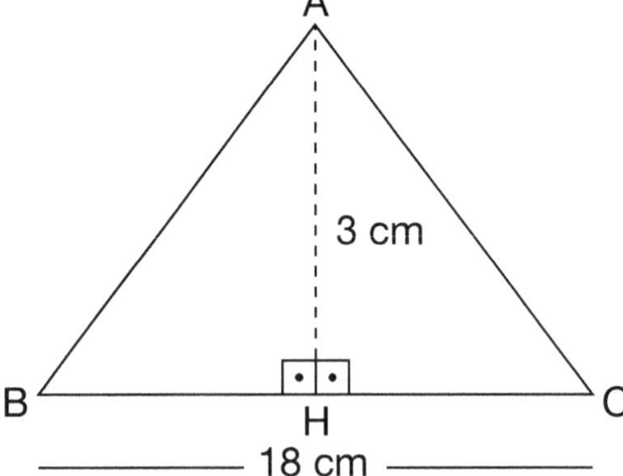

What is the area of △ BAC?

A) 18 cm^2

B) 27 cm^2

C) 30 cm^2

D) 36 cm^2

SAMPLE TEST

25. Find the perimeter of the following triangle

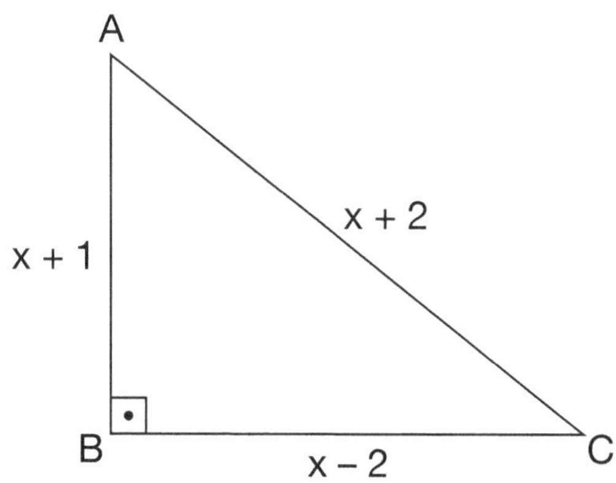

A) 3x

B) 3x – 1

C) 3x + 1

D) 3x + 2

26.

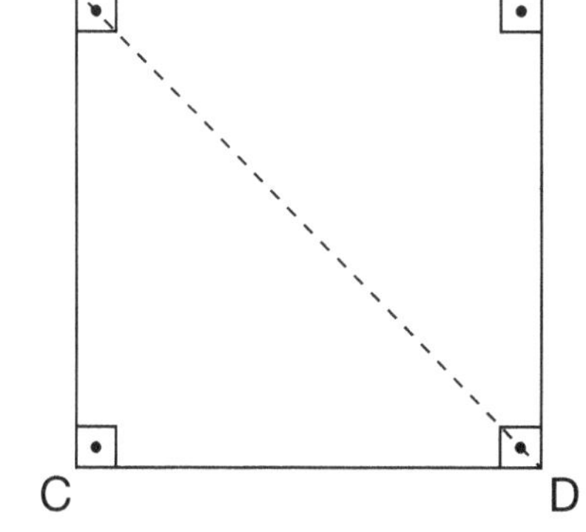

What is the area of the square above if the length of AD is $6\sqrt{2}$?

A) 16

B) 25

C) 36

D) 47

27. The angles of a triangle are in the ratio of 4: 5: 9. What is the degree measure of the smallest angle?

A) 40°

B) 30°

C) 25°

D) 20°

28.

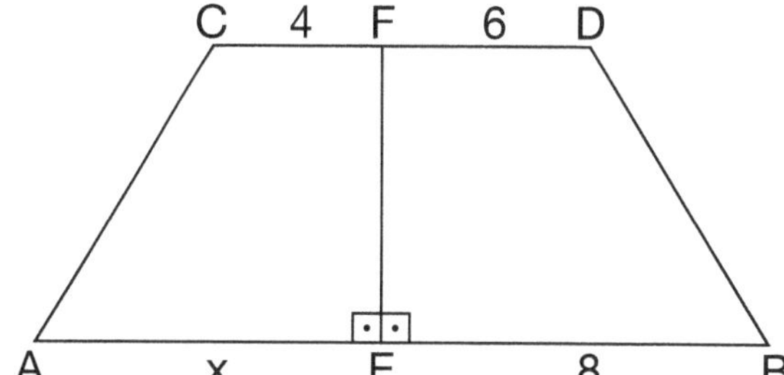

In the above figure, if the area of AEFC is equal to the area of EBDF,

what is the value of x?

A) 6

B) 8

C) 10

D) 12

29. The perimeter of a rectangle is P = 2L + 2W. What is the width in terms of the length and perimeter?

A) $\frac{P}{2} + L$

B) $\frac{P}{2} - L$

C) P + L

D) P – 2L

30. In the following figure, O is the center of the circle. What is the measure of x?

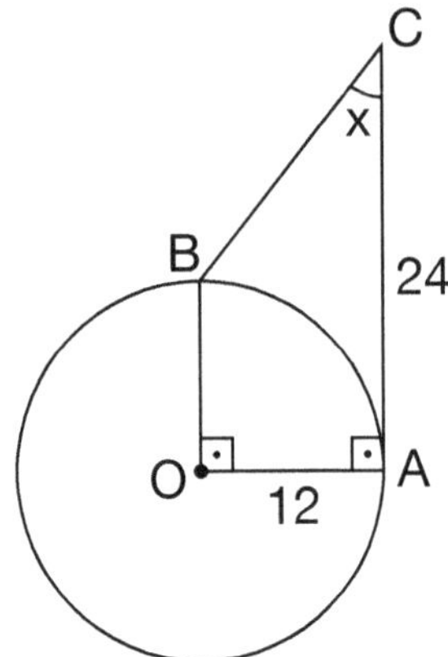

A) 15°

B) 25°

C) 30°

D) 45°

31. If the center of a circle is at (3,6), and the radius of the circle is 4, what is the equation of that circle?

A) $(x - 3)^2 + (y - 6)^2 = 4$

B) $(x - 3)^2 + (y - 6)^2 = 8$

C) $(x - 3)^2 + (y - 6)^2 = 16$

D) $(x - 6)^2 + (y - 3)^2 = 4$

32.

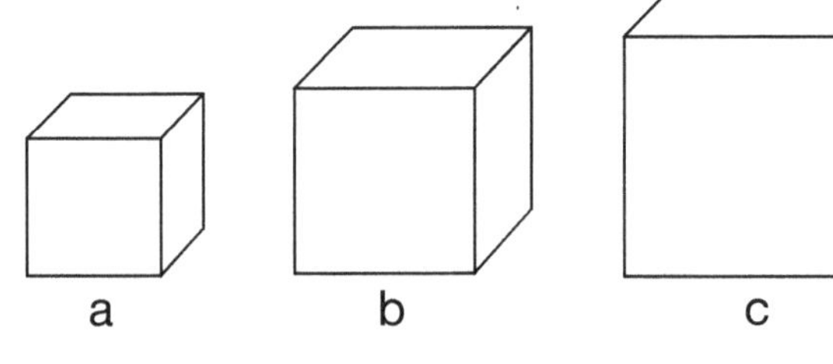

There are three cubes given above. Their side length is given under the cubes. 'I' is a, 'II' is b, 'III' is c. Their volumes are V_a, V_b, and V_c.

$$c = 2b = 3a$$

If $\frac{1}{V_a} + \frac{1}{V_b} + \frac{1}{V_c} = \frac{1}{6}$, then find the volume of V_c.

A) 216

B) 125

C) 64

D) 27

33. The radius of a cylinder is increased by 25% and its height is decreased by 20%. What is the effect on the volume of the cylinder?

A) It is decreased by 50%

B) It is decreased by 25%

C) It is increased by 25%

D) It is increased by 50%

34. In a right triangle, the cosine of angle B is $\frac{3}{5}$ and the sine of angle B is $\frac{4}{5}$.
What is the ratio of the longest side to the shortest side?

A) $\frac{4}{3}$

B) $\frac{4}{5}$

C) $\frac{3}{4}$

D) $\frac{5}{3}$

LOCICAL REASONING

Propositional Logic:

- This is the branch of logic dealing with propositions (declarative statements that can be true or false) and logical connectives like AND, OR and NOT.
- Example : "It is raining" is a proposition which can be true or false.

Logical Connectives:

- **Conjunction (AND, $\wedge$):** True only if both propositions are true.
- **Disjunction (OR, $\vee$):** True if at least one of the propositions is true.
- **Negation (NOT, $\neg$):** True if the proposition is false, and false if the proposition is true.

Conditional Statements:

- Formed with "if...then..." (symbolized as $p \longrightarrow q$).
- Example : "If it is raining (p), then the ground is wet (q)."

Biconditional Statements:

- Formed with "If and only if" (symbolized as $p \longleftrightarrow q$).
- Exampl"eA figure is a square (p) if and only if it has four equal sides (q)."

Converse: The converse of the statement "If P, then Q" is "If Q, then P."

Example: If our original statement was "If a shape is a square, then it has four equal sides," the converse would be "If a shape has four equal sides, then it is a square."

Inverse: The inverse of the statement "If P, then Q" is "If not P, then not Q."

Example: "If a shape is not a square, then it does not have four equal sides."

Contrapositive: The contrapositive of the statement "If P, then Q" is "If not Q, then not P."

Example: "If a shape does not have four equal sides, then it is not a square."

LOCICAL REASONING

Here are some important things to note:

- If the original conditional statement is true, then its contrapositive is also true.
- If the converse is true, then its inverse is also true.
- However, the truth of the original statement does not guarantee the truth of its converse or inverse.

Example: "If an angle measures 90 degrees, then it is a right angle."

- Converse: "If an angle is a right angle, then it measures 90 degrees."
- Inverse: "If an angle does not measure 90 degrees, then it is not a right angle."
- Contrapositive: "If an angle is not a right angle, then it does not measure 90 degrees."

Truth Tables:

- Tables used to determine the truth value of a compound statement based on the truth values of the individual propositions.
- Example : Truth table for $p \wedge q$:

p	q	$p \wedge q$
T	T	T
T	F	F
F	T	F
F	F	F

Rules of Inference:

- Rules used to derive conclusions from premises.
- Example : If $p \longrightarrow q$ and p are both true, then q is true.

Counterexamples:

- Examples used to prove a statement or proposition false.
- Example : To disprove "All birds can fly." a counterexample could be "Penguins are birds, but they cannot fly."

LOCICAL REASONING

1. Which of the following is a proposition?

A) Go to the store!

B) How are you?

C) It is raining.

D) Read this book.

2. What is the negation of the statement "It is cold"?

A) It is not cold

B) It is warm

C) It is hot

D) It is chilly

3. In propositional logic, what does the symbol $\wedge$ represent?

A) OR

B) AND

C) NOT

D) IF

4. Which of the following represents a biconditional statement?

A) $p \longrightarrow q$

B) $p \vee q$

C) $p \wedge q$

D) $p \longleftrightarrow q$

5. If p represents "It is raining" and q represents "The ground is wet", how is the statement "It is raining or the ground is wet" represented symbolically?

A) $p \longrightarrow q$

B) $p \wedge q$

C) $p \vee q$

D) $p \longleftrightarrow q$

LOCICAL REASONING

Write the converse, the inverse, and the contrapositive of the following conditionals

6. If a bird can sing, then it is not a penguin.

Converse:

Inverse:

Contrapositive:

7. If a car is electric, then it does not use gasoline.

Converse:

Inverse:

Contrapositive:

8. If a shape has four equal sides, then it is a square.

Converse:

Inverse:

Contrapositive:

9. If a planet orbits the sun, then it is not outside our solar system.

Converse:

Inverse:

Contrapositive:

10. If a number is even, then it is not prime.

Converse:

Inverse:

Contrapositive:

ALGEBRAIC PROOFS

Algebraic Proof: An algebraic proof is a type of justification used to prove the validity of a mathematical statement using algebraic properties and operations.

In such a proof, the algebraic properties, like the distributive property, commutative property, associative property, etc..,are applied systematically to derive a conclusion.

	Definition	Example
Reflexive Property	For every number a, a = a.	For any real number x, x = x.
Symmetric Property	For all numbers a and b, if a = b then b = a.	If 2x = 6, then 6 = 2x.
Transitive Property	For all numbers a, b and c, if a = b and b = c then a = c	If y = 3x and 3x = 12, then y = 12.
Addition Property	For all numbers a, b and c, if a = b then a + c = b + c	If x = 4, then x + 3 = 7.
Subtraction Property	For all numbers a, b and c, if a = b then a − c = b − c	If m = 5, then m − 2 = 3.
Multiplication Property	For all numbers a, b and c, if a = b then a x c = b x c.	If z = 3, then 2z = 6.
Division Property	For all numbers a, b and c (with c ≠ 0), if a = b then a ÷ c = b ÷ c	If p = 8, then p ÷ 2 = 4.
Substitution Property	For all numbers a and b, if a = b then a may be replaced by b in any equation or expression.	Given ω = 4x, in the expression ω + 3, ω can be replaced by 4x, yielding 4x + 3.
Distributive Property	For all numbers a, b and c, a(b + c) = ab + ac.	For 3(2x + 4), distribute the 3 to get 6x + 12.

ALGEBRAIC PROOFS TEST

1. **Given :** $7x - 3(2x + 4) = 10$.

Prove : $x = 22$

2. **Given :** $3x + 5(x - 3) = 17$.

Prove : $x = 4$

3. **Given :** $2x + 12 = 3x + 18$.

Prove : $x = -6$.

4. **Given:** $4x + 14 = 5x + 20$.

Prove : $x = -6$.

GEOMETRIC PROOFS

Parallel Lines

Parallel lines are two or more lines that are an equal distance apart everywhere and they will never meet, no matter how long you extend them.

Proofs involving parallel lines:

1. **Corresponding Angles Postulate:** If two parallel lines are cut by a transversal, then the pairs of corresponding angles are congruent.
2. **Alternate Interior Angles Postulate:** If two parallel lines are cut by a transversal, then the pairs of alternate interior angles are congruent.
3. **Alternate Exterior Angles Postulate:** If two parallel lines are cut by a transversal, then the pairs of alternate exterior angles are congruent.
4. **Consecutive Interior Angles Postulate:** If two parallel lines are cut by a transversal, then the pairs of consecutive interior angles are supplementary (their measures add up to 180 degrees).

Perpendicular Lines

Perpendicular lines intersect at right angles (90 degrees). The symbol ⊥ is used to denote perpendicular lines. For example, $l \perp m$ means line l is perpendicular to line m.

Proofs involving perpendicular lines:

1. **Perpendicular Lines Form Right Angles:** If two lines are perpendicular, then they form right angles.
2. **Converse:** If two lines form right angles, then they are perpendicular.
3. **Perpendicular Transversal Theorem:** In a plane, if a line is perpendicular to one of two parallel lines, it is also perpendicular to the other.

Example:

Given: Line l is parallel to line m. Line t is perpendicular to line l.

Prove: Line t is perpendicular to line m.

Solution:

1. Line l is parallel to line m. (Given)
2. Line t is perpendicular to line l. (Given)
3. Line t forms a right angle with line l. (From 2, because perpendicular lines form right angles)
4. Line t also forms a right angle with line m. (Using the Perpendicular Transversal Theorem)
5. Therefore, line t is perpendicular to line m.

GEOMETRIC PROOFS

Angle Addition Postulate : If point T is the interior of ∠ UVD, then the measure of ∠ UVT added to the measure of ∠ TVD equals the measure of ∠ UVD.

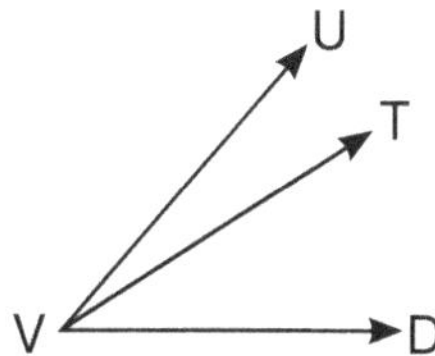

Supplement Theorem : If two angles form a linear pair, they are supplementary angles. This means the sum of their measures is 180°.

If ∠ 1 and ∠ 2 form a linear pair, then m ∠ 1 + m ∠ 2 = 180.

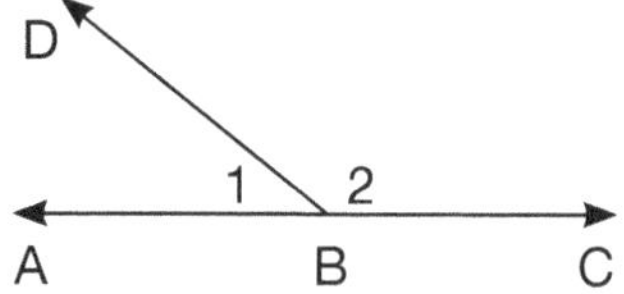

Complement Theorem : If the noncommon sides of two adjacent angles form a right angle, these angles are complementary angles, meaning the sum of their measures is 90°.

If $\overleftrightarrow{GF} \perp \overleftrightarrow{GH}$, then m ∠ 3 + m ∠ 4 = 90.

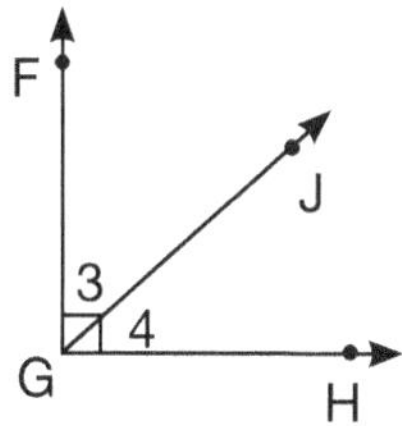

Segment Addition Postulate

A B C

If point B is between points A and C, then the length of segment AB plus the length of segment BC is equal to the length of segment AC.

Mathematically, it can be expressed as:

AB + BC = AC

when B is between A and C.

GEOMETRIC PROOFS TEST

1. **Given :** B is the midpoint of $\overline{AC}$.

C is the midpoint of $\overline{BD}$

Prove : $\overline{AC} = \overline{BD}$

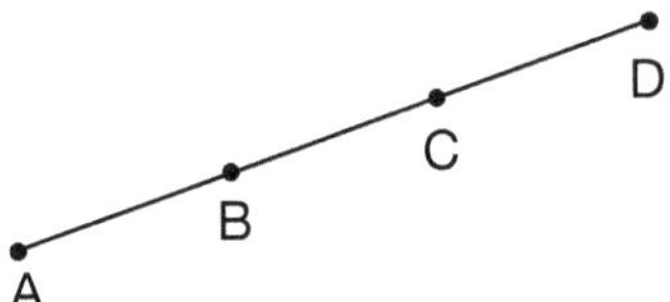

2. **Given :** ∠ 2 and ∠ 3 are supplementary.

Prove : A || B

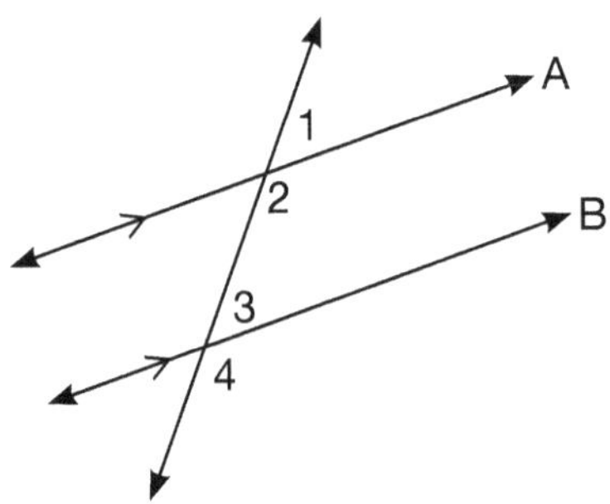

3. **Given :** a ll b, c ⊥ a

Prove : c ⊥ b

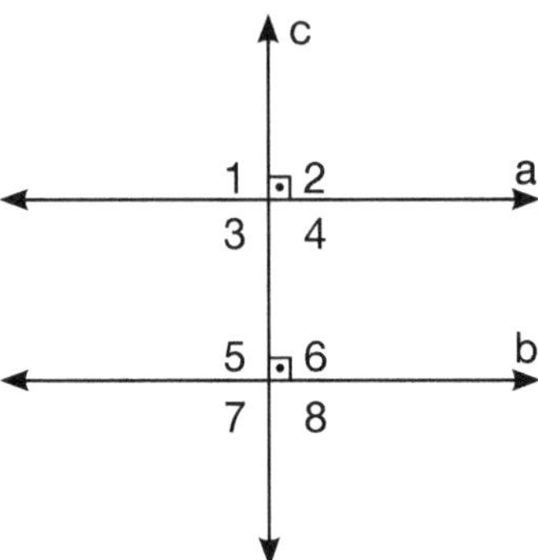

4. **Given :** ∠ 3 and ∠ 4 are vertical angles.

Prove : ∠ 3 ≅ ∠ 4

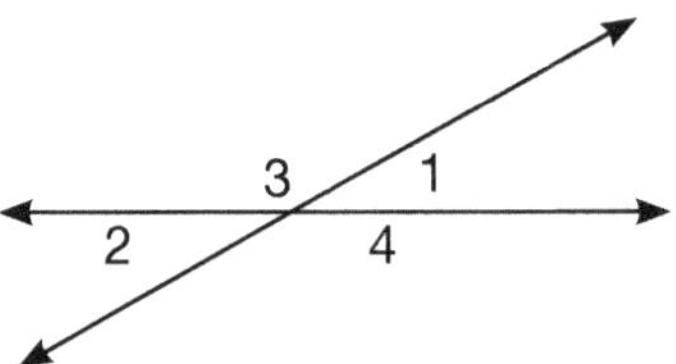

GEOMETRIC PROOFS TEST

5. **Given :** Two angles, $\angle$ A and $\angle$ B, are complementary.

Prove : m $\angle$ A + m $\angle$ B = 90°

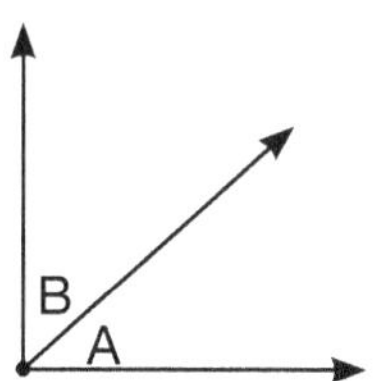

6. **Given :** Two angles, $\angle$ A and $\angle$ B, are supplementary.

Prove : m $\angle$ A + m $\angle$ B = 180°

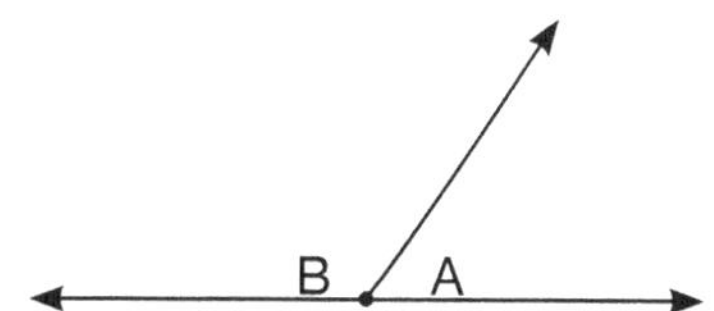

ANGLE RELATIONSHIPS

Point: A point is a location represented by a dot. It has no length, no width, or size. It only has a position.

Line: A line is made of a set of points which is extended infinitely in both directions

Ray: A ray is a part of a line that starts at one point and continues infinitely in only one direction

Line Segment: A line segment is part of a line with an endpoint on both sides

Angle: An angle is created when two rays connect at a common point. The common endpoint (Y), called a vertex.

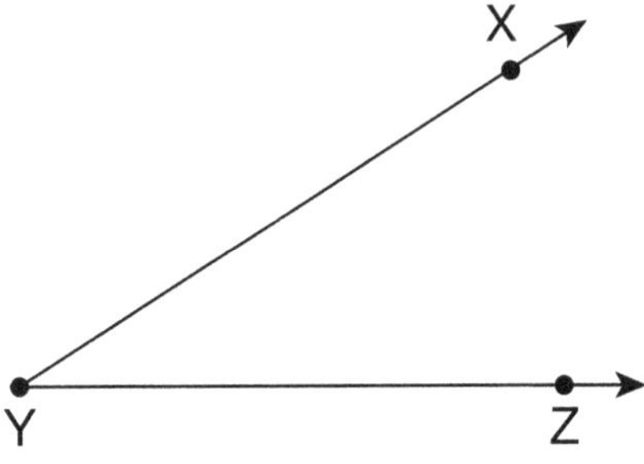

Acute Angle: An angle that is less than 90° but greater than 0°

0° < Acute Angle < 90°

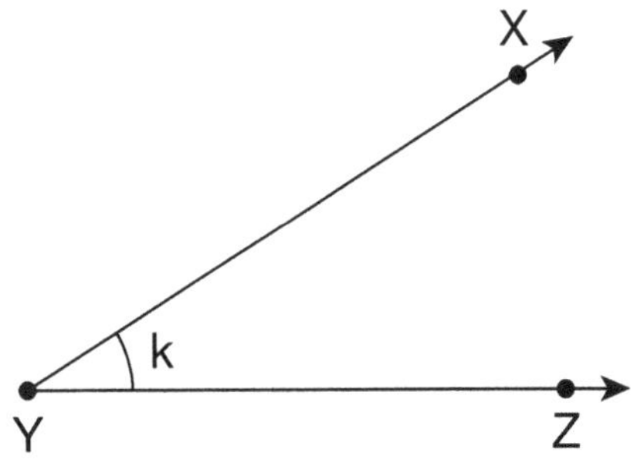

$0° < k < 90°$

ANGLE RELATIONSHIPS

Right Angle: An angle that is exactly 90°

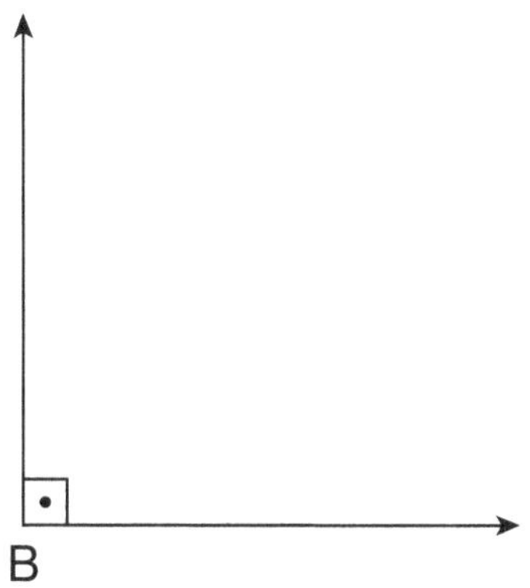

Obtuse Angle: An angle that is greater than 90° but less than 180°

90° < Obtuse Angle < 180°

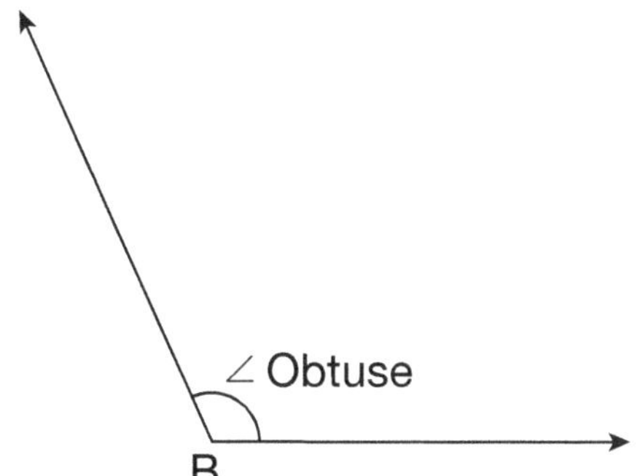

$90° < m\angle B < 180°$

Complementary Angles: Complementary angles are two angles whose measures add up to 90°

✓ In the figure below, ∠x and ∠y are complementary angles.

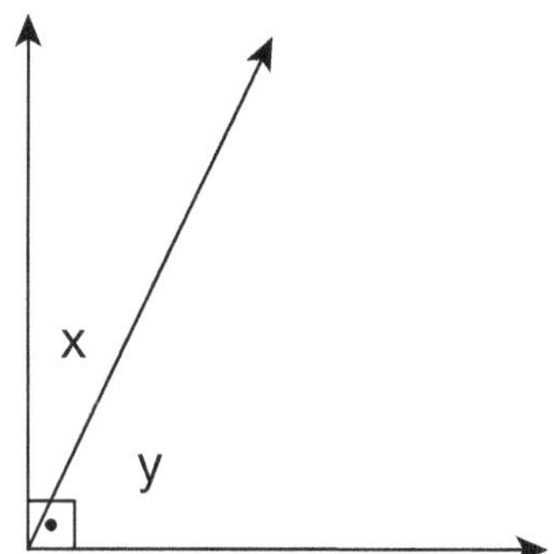

$m\angle x + m\angle y = 90°$

ANGLE RELATIONSHIPS

Supplementary Angles: Supplementary angles are two angles whose measures add up to 180°

✓ In the figure below, $\angle x$ and $\angle y$ are supplementary angles.

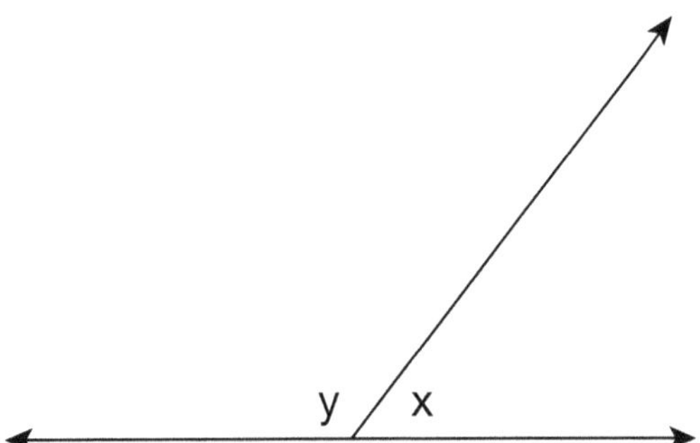

$m\angle x + m\angle y = 180°$

Vertical Angles: Vertical Angles are the angles formed by two intersecting lines. Vertical angles are always congruent and equal to each other.

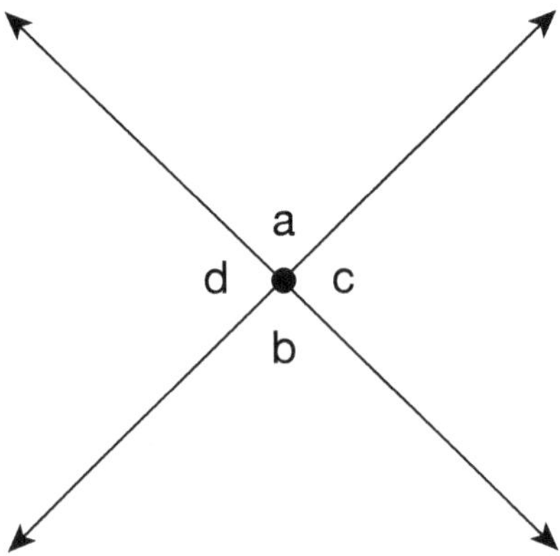

✓ In the figure, $\angle a$ and $\angle b$ are equal to each other and $\angle c$ and $\angle d$ are equal to each other.

✓ In the figure, $\angle a$ and $\angle b$ are congruent each other and $\angle c$ and $\angle d$ are congruent each other.

Straight Angle: An angle that is exactly 180°

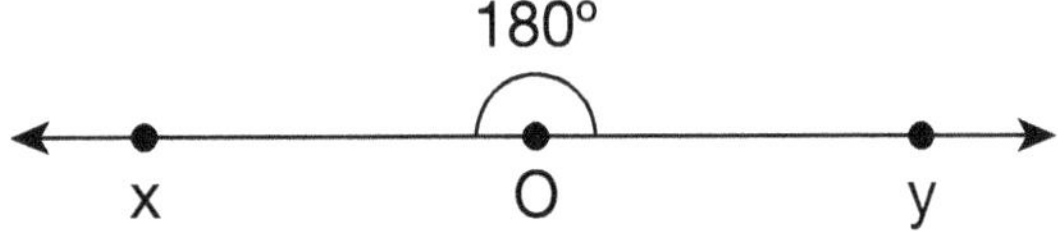

ANGLE RELATIONSHIPS TEST

1. What is the value of x in the diagram below?

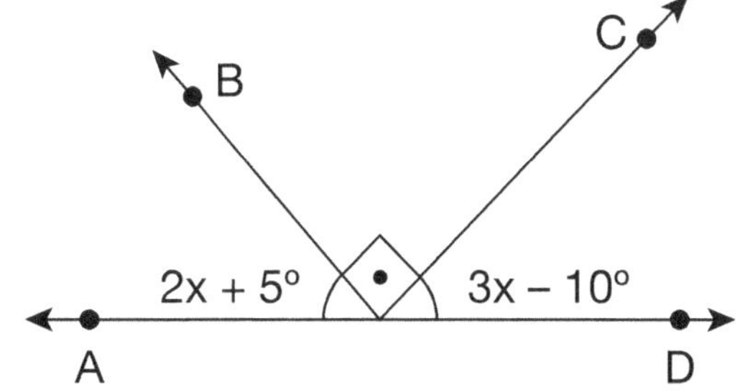

A) 10°

B) 15°

C) 17°

D) 19°

2. What is the value of x in the diagram below?

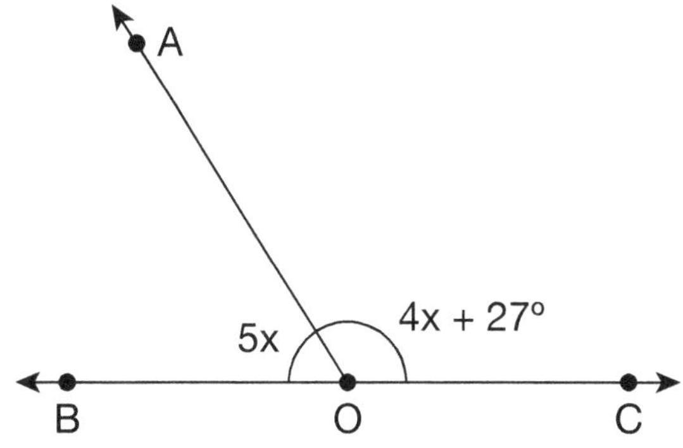

A) 15°

B) 17°

C) 21°

D) 33°

3. What is the value of x in the diagram below?

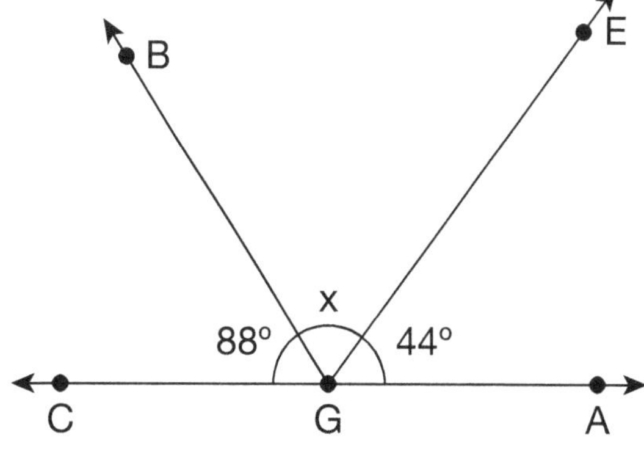

A) 12°

B) 18°

C) 24°

D) 48°

4. What is the value of x in the diagram below?

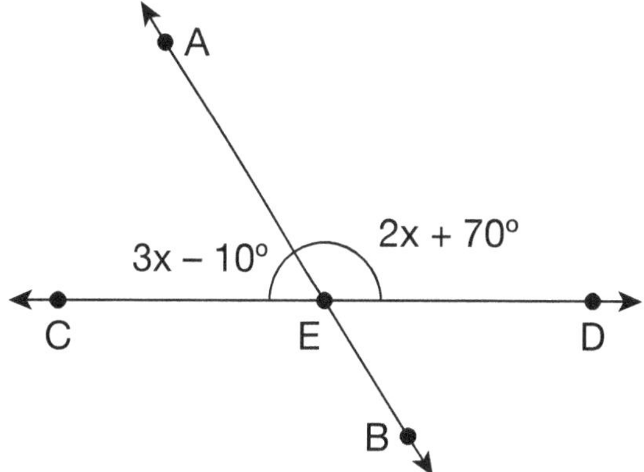

A) 9°

B) 12°

C) 24°

D) 36°

5. What is the value of x in the diagram below?

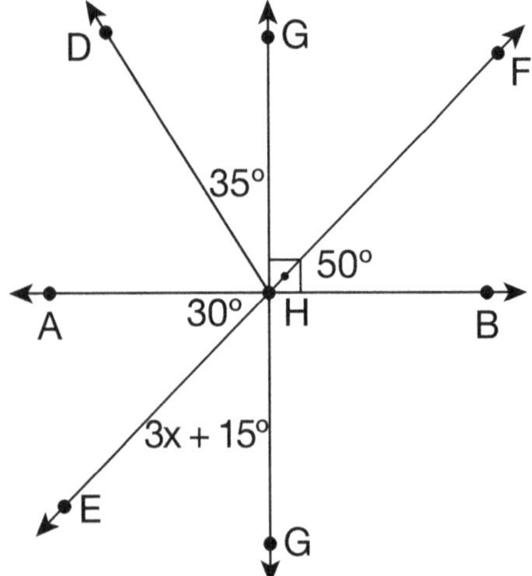

A) 15°

B) 20°

C) 30°

D) 45°

6. What is the value of x in the diagram below? Given: ℓ // m

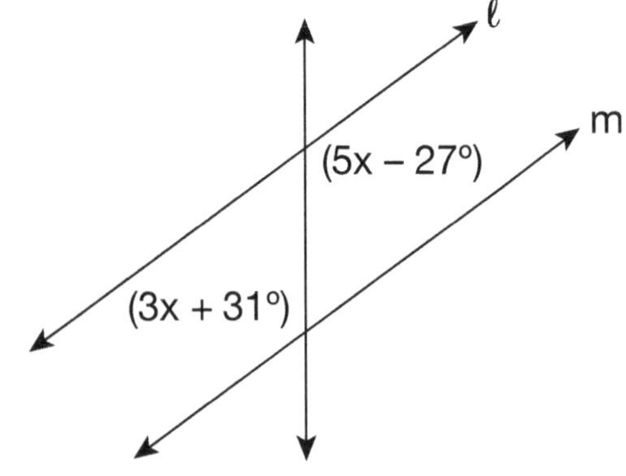

A) 19°

B) 29°

C) 39°

D) 49°

7. What is the value of x in the diagram below?

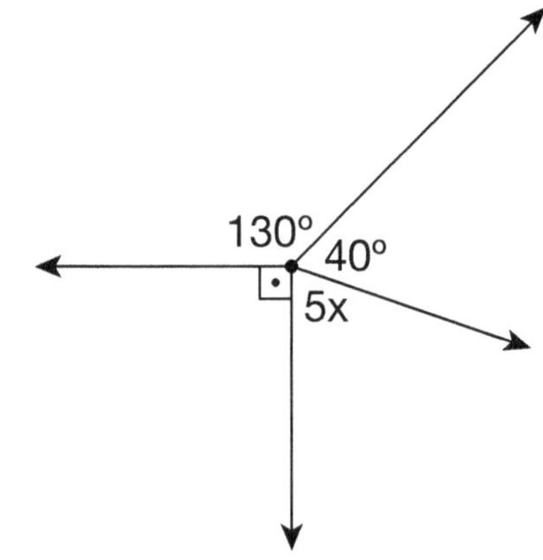

A) 5°

B) 10°

C) 20°

D) 25°

8. What is the value of x in the diagram below?

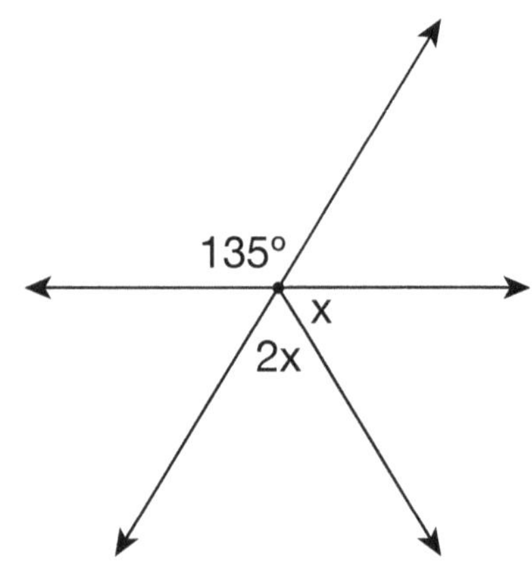

A) 25°

B) 30°

C) 35°

D) 45°

ANGLE RELATIONSHIPS TEST

9. What is the value of x in the diagram below?

Given: AB // DE.

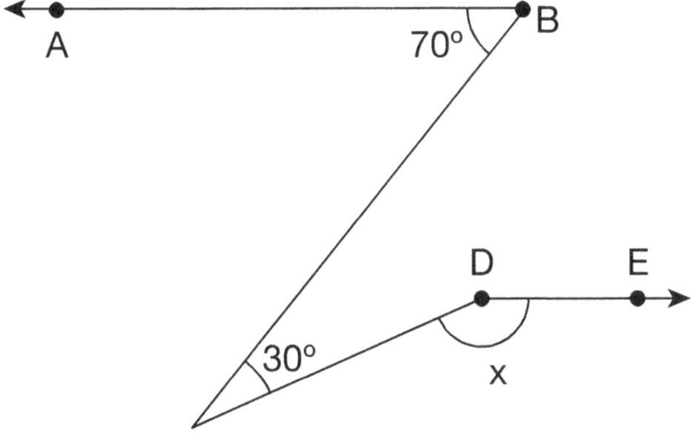

A) 120°

B) 140°

C) 150°

D) 170°

10. What is the value of x in the diagram below?

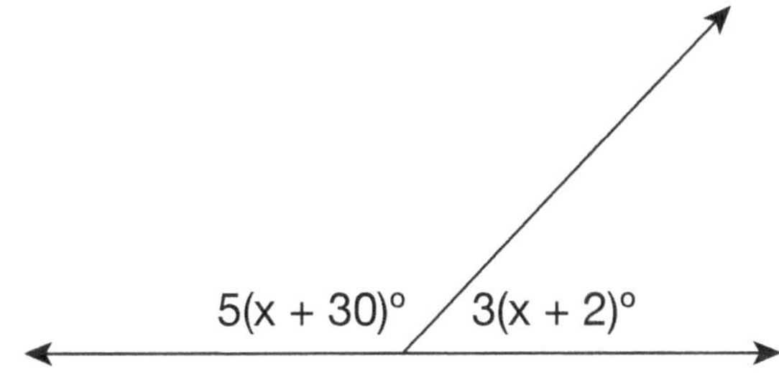

A) 3°

B) 5°

C) 8°

D) 10°

11. What is the value of x+y given the information in the diagram below?

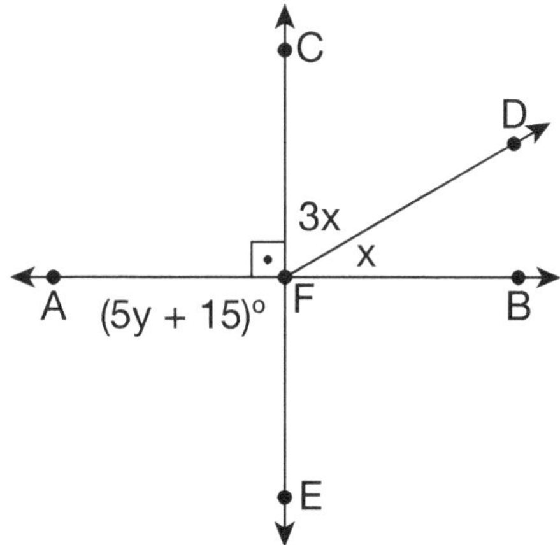

A) 18°

B) 22.5°

C) 37.5°

D) 45°

12. What is the value of x in the diagram below?

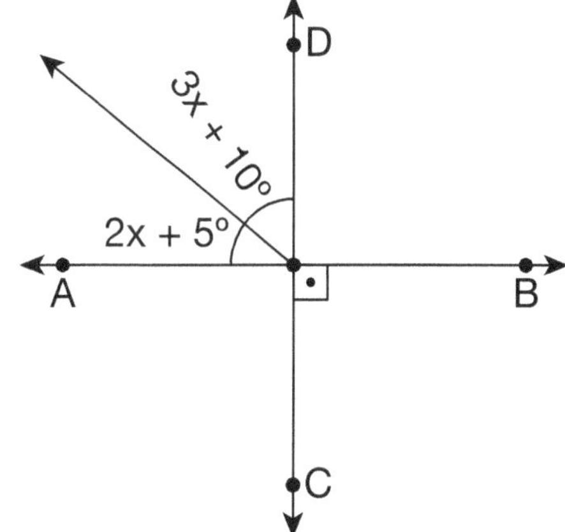

A) 10°

B) 15°

C) 25°

D) 35°

INTERIOR AND EXTERIOR ANGLES

Given: k//l

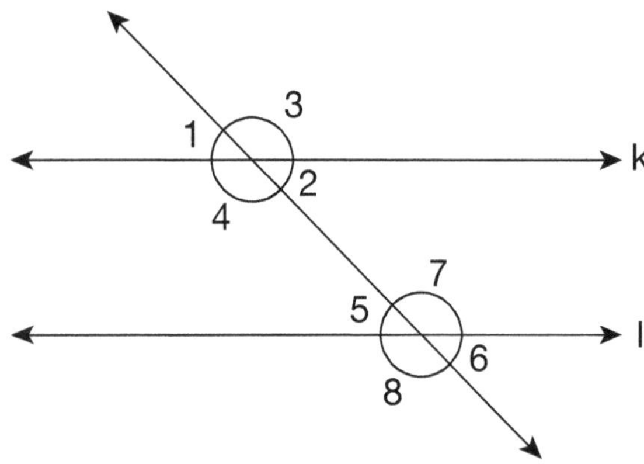

Interior Angles: The angles that are formed inside two parallel lines that are cut by a transversal. In the above figure, $\angle 2$, $\angle 4$, $\angle 5$, and $\angle 7$ are interior angles.

Exterior Angles: The angles that are formed outside of two parallel lines that are cut by a transversal. In the above figure, $\angle 1$, $\angle 3$, $\angle 6$, and $\angle 8$ are exterior angles.

Alternate Interior Angles: The angles that are on opposite sides of a transversal. Alternate interior angles are congruent. In the above figure, $\angle 2 \cong \angle 5$, and $\angle 4 \cong \angle 7$ are alternate interior angles.

Alternate Exterior Angles: The angles that are on opposite sides of a transversal. Alternate exterior angles are congruent. In the above figure, $\angle 1 \cong \angle 6$, and $\angle 3 \cong \angle 8$ are alternate exterior angles.

Corresponding Angles: The angles that are formed when two parallel lines are intersected by the transversal. Corresponding angles are congruent. In the above figure, $\angle 1 \cong \angle 5$, and $\angle 3 \cong \angle 7$ are corresponding angles.

INTERIOR AND EXTERIOR ANGLES TEST

Use the following figure to answer questions 1 through 8.

Given: m // n

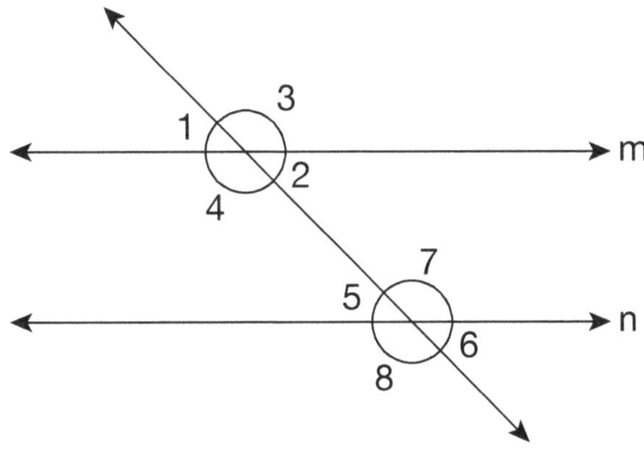

1) If m $\angle 6 = 70^\circ$, what is m $\angle 1$? ______________________________

2) If m $\angle 7 = 110^\circ$, what is m $\angle 3$? ______________________________

3) List all pairs of alternate interior angles: ______________________________

4) List all pairs of alternate exterior angles: ______________________________

5) List all pairs of corresponding angles: ______________________________

6) List all pairs of supplementary angles: ______________________________

7) If m $\angle 5 = 75^\circ$, what is m $\angle 7$? ______________________________

8) If m $\angle 8 = 125^\circ$, what is m $\angle 4$? ______________________________

SEGMENT ADDITION POSTULATE

Proof: A method of determining whether a statement is true or false with logical reasons used to confirm an idea.

Theorem: A result that has been proved to be true using other theorems or postulates.

Postulate: Is a statement that is accepted true without proof.

Midpoint: The point that bisects a segment.

Example:

If point Y bisects $\overline{XZ}$, then XY = YZ

Segment Addition Postulate:

If y is a point on line segment $\overline{XZ}$, then XY + YZ = XZ

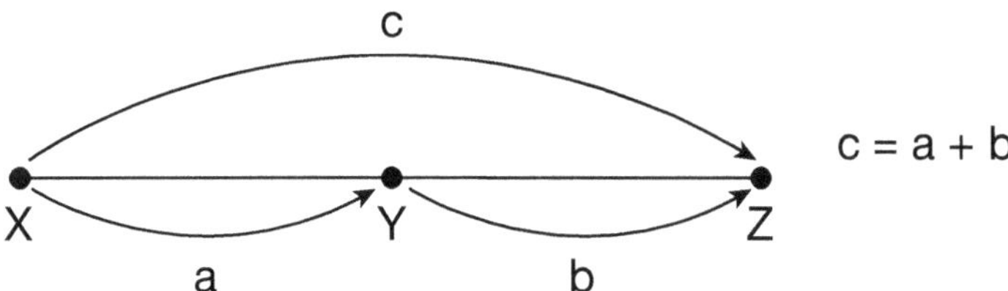

c = a + b

SEGMENT ADDITION POSTULATE TEST

1. If B is between A and C, AC = 15, find the value of x.

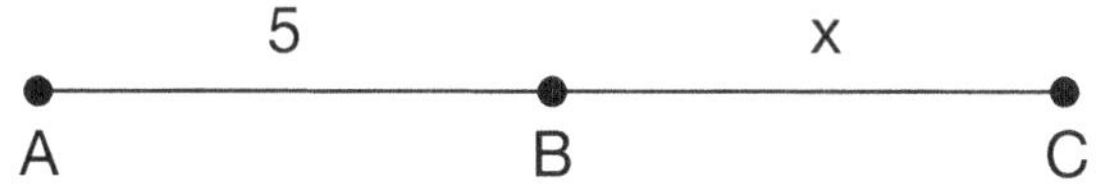

A) 5

B) 10

C) 15

D) 20

2. If B is between A and C, AC = 14, find the value of x.

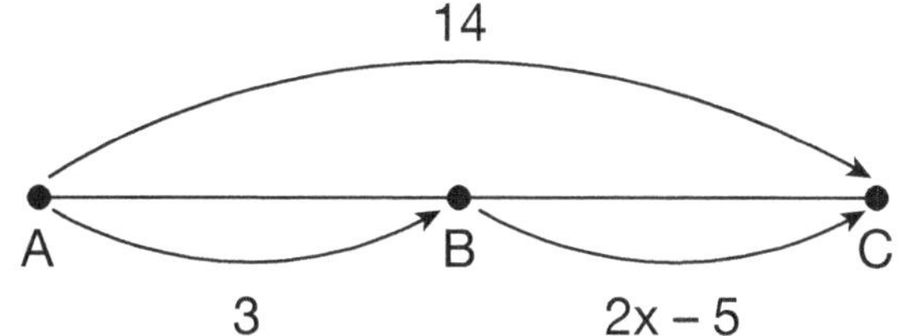

A) 4

B) 8

C) 12

D) 16

3. If C is between A and B and AB = 20, AC = 2x + 3, and BC = 3x – 8, find the value of x.

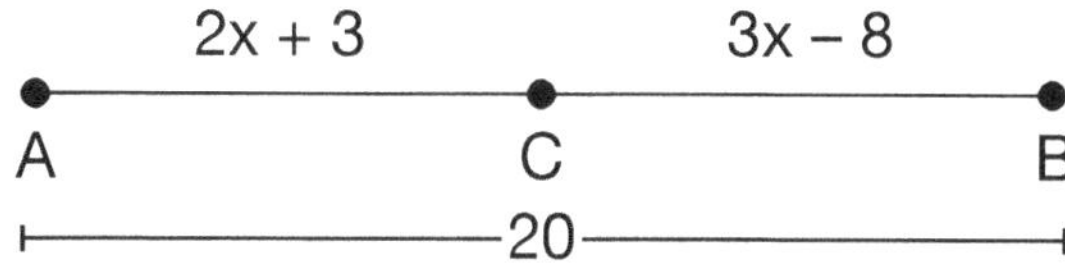

A) 5

B) 10

C) 15

D) 20

4. If B is the midpoint of segment AC and AB = 2x – 12, BC = 5x – 42, find the value of x.

A) 9

B) 10

C) 12

D) 15

5. If B is the midpoint of segment AC and AB = 5x – 15 and BC = 3x + 12, find the value of x.

A) 5

B) 10

C) 11.5

D) 13.5

6. If AD = 3x + 40, BC = 8, BD = x + 10, and AC = x + 45, find the value of x.

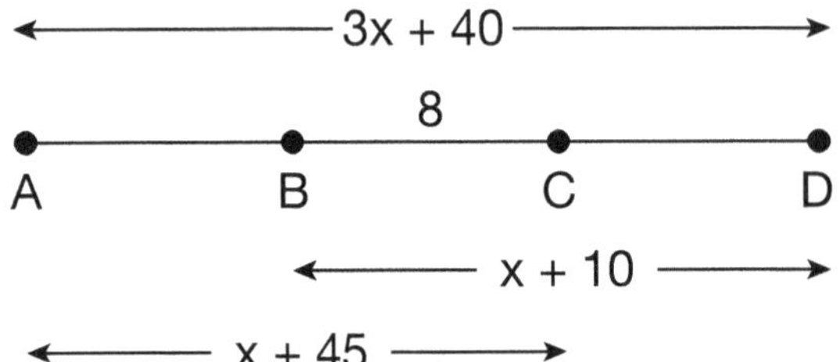

A) 5

B) 7

C) 9

D) 12

ANGLE BISECTORS

An angle bisector is a line that splits an angle into two congruent angles.

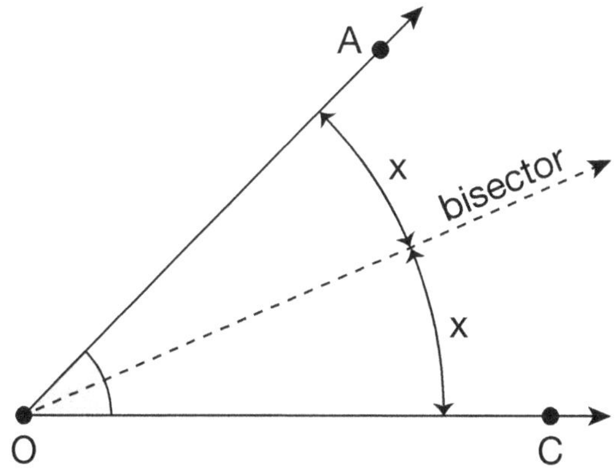

If $\overrightarrow{OC}$ is the bisector of $\angle AOC$, then $\angle AOB \cong \angle BOG$

Example:

EF is the angle bisector of m $\angle DEG$, m $\angle DEF = 2x + 10°$ and m $\angle FEG = 3x + 5°$, then find x.

Solution:

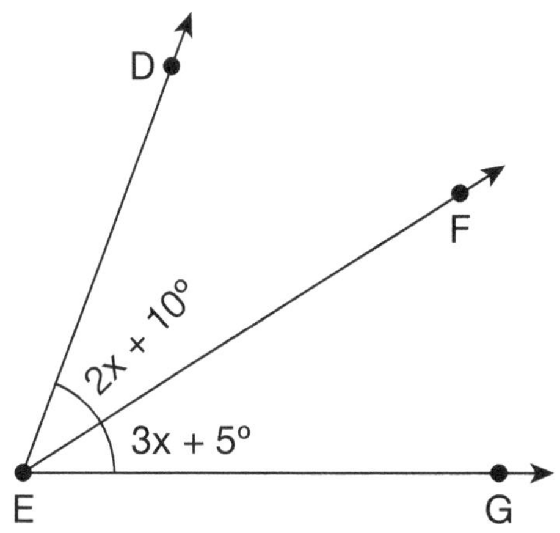

Since EF is bisector of m $\angle DEG$, then m $\angle DEF$ = m $\angle FEG$

$$2x + 10° = 3x + 5$$

$$-2x \qquad -2x$$

$$10 = x + 5, \text{ then } x = 5$$

ANGLE BISECTORS TEST

1. WY is the angle bisector of m $\angle$ XWZ, m $\angle$ XWY = 2x + 16°, and m $\angle$ YWZ = 3x – 8°. Find x.

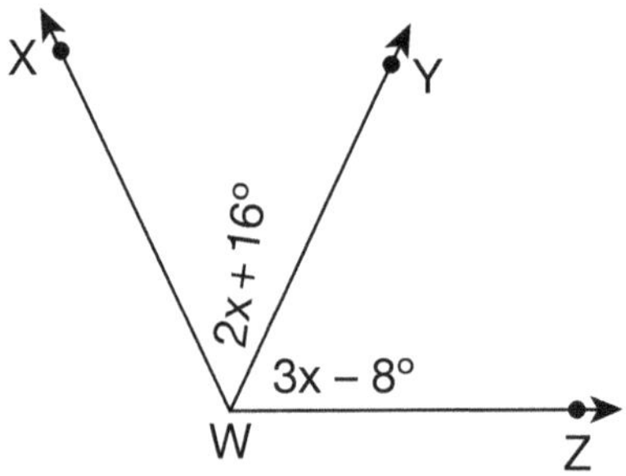

A) 12°

B) 18°

C) 24°

D) 48°

2. EF is the angle bisector of m $\angle$ DEG, m $\angle$ DEF = 4x – 12°, and m $\angle$ FEG = 3x + 24°.Find x.

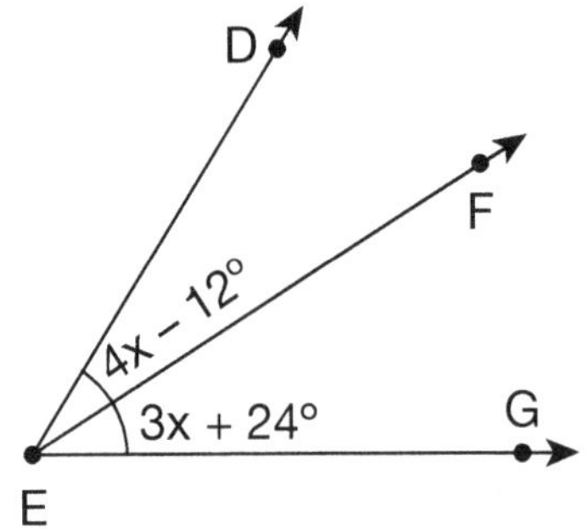

A) 12°

B) 18°

C) 32°

D) 36°

3. JK is the angle bisector of m $\angle$ LJM

m $\angle$ LJK = $\frac{x}{2} - 10^\circ$, and

m $\angle$ KJM = $\frac{3x}{2} - 24^\circ$. Find x.

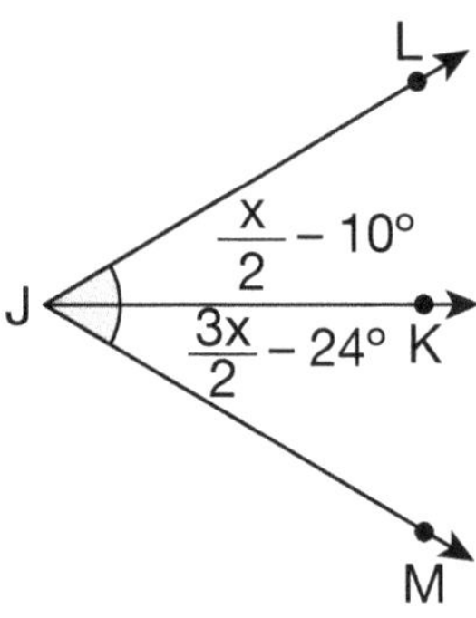

A) 14°

B) 28°

C) 32°

D) 48°

4. m is the bisector of $\overline{AB}$, AO = 2x – 30, and

OB = $\frac{x-2}{3} + 15$. Find x.

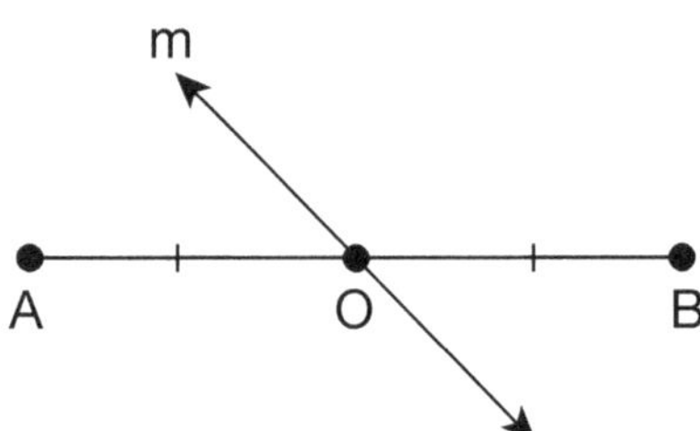

A) 12.5°

B) 18.6°

C) 26.6°

D) 34.5°

American Math Academy

CLASSIFYING TRIANGLES

Equilateral triangles

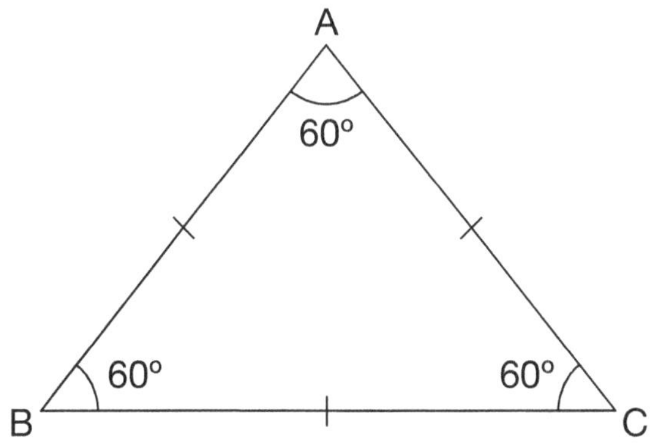

✓ All the three sides are equal

✓ All the three angles are equal to 60°

Isosceles triangle

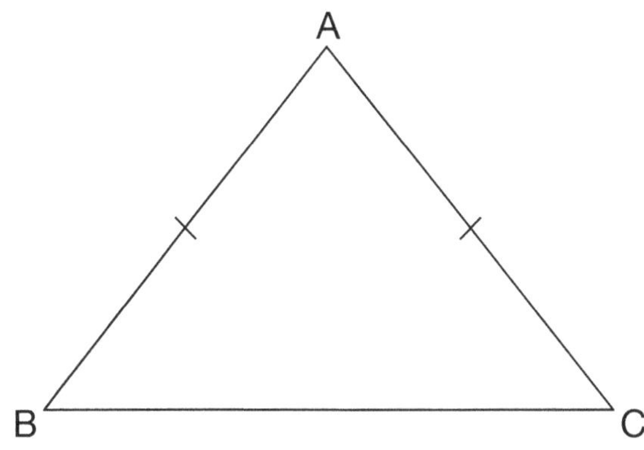

✓ At least any two sides are equal

✓ Any two angles are equal

Scalene triangle

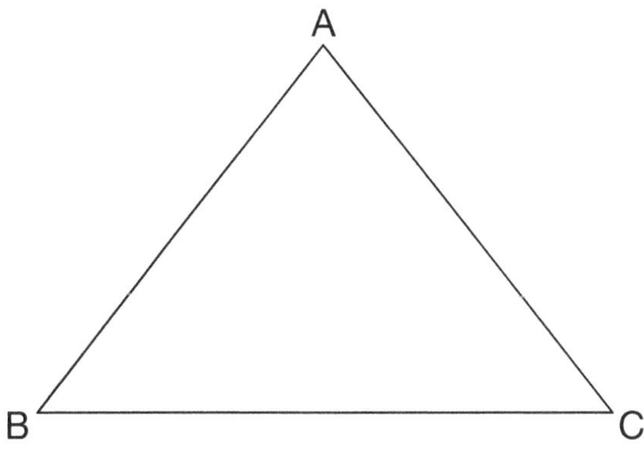

✓ All three sides have different lengths

✓ All three angles are unequal

CLASSIFYING TRIANGLES

Acute triangle

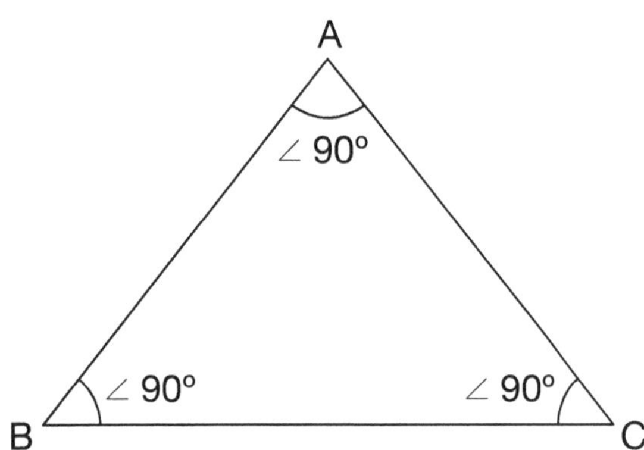

✓ All angles are less than 90°

Right triangle

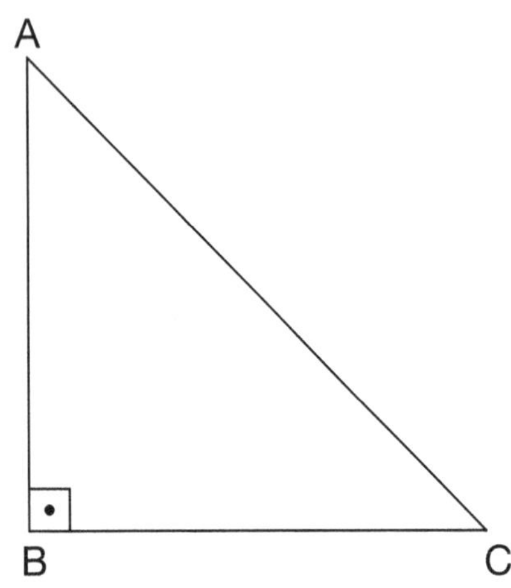

✓ One angle is exactly 90°

Obtuse triangle

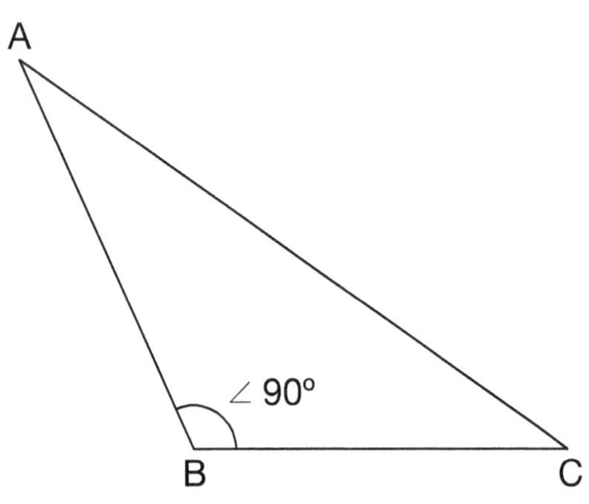

✓ One angle is greater than 90° but less than 180°

$90° < m\angle B < 180°$

CLASSIFYING TRIANGLES TEST

1. $\overrightarrow{AD}$ bisect $\angle$ BAC. Find the value of x.

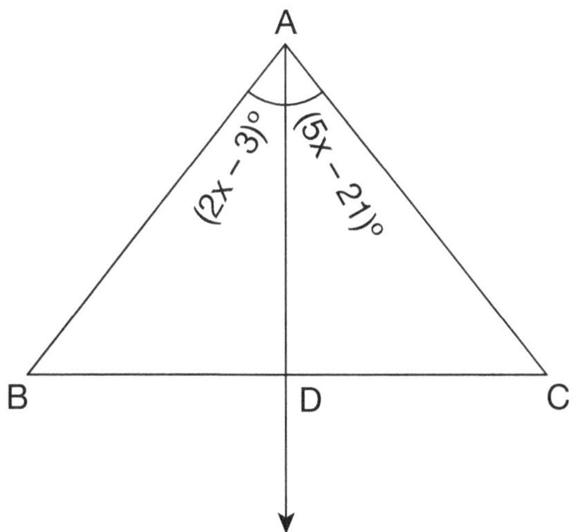

A) 3°

B) 6°

C) 9°

D) 12°

2. Find the value of x in the isosceles triangle below.

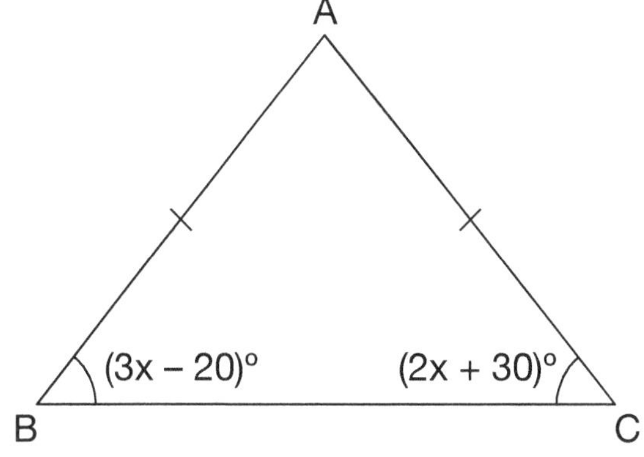

A) 15°

B) 25°

C) 35°

D) 50°

3. Find the value of x in the triangle below.

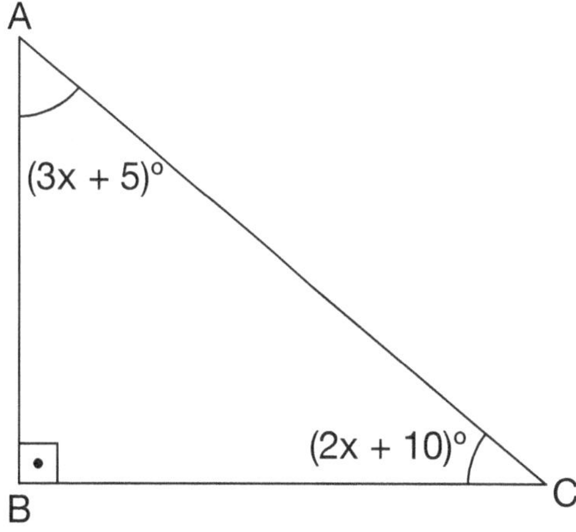

A) 5°

B) 10°

C) 15°

D) 20°

4. Find the value of x in the isosceles triangle below.

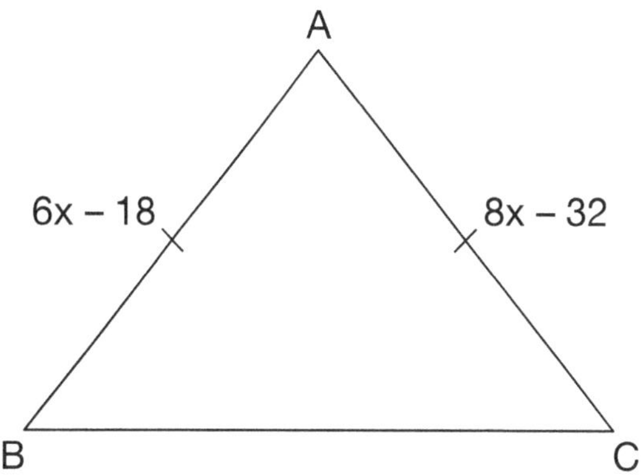

A) 7°

B) 14°

C) 17°

D) 21°

5. Find the value of x in the triangle below.

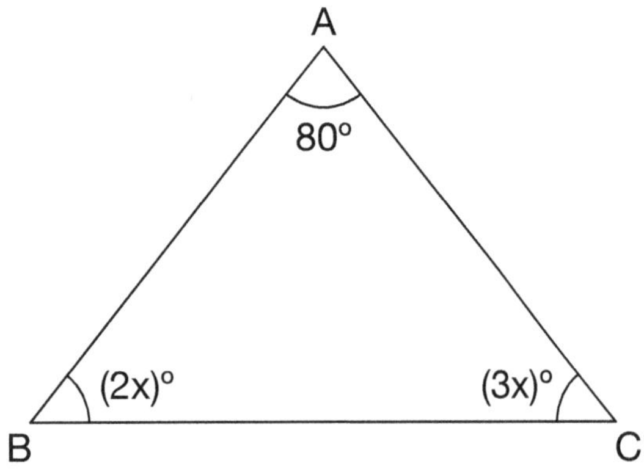

A) 5°

B) 10°

C) 15°

D) 20°

INTERIOR AND EXTERIOR TRIANGLES

Interior angles: The sum of the interior angles of a triangle is equal to 180.

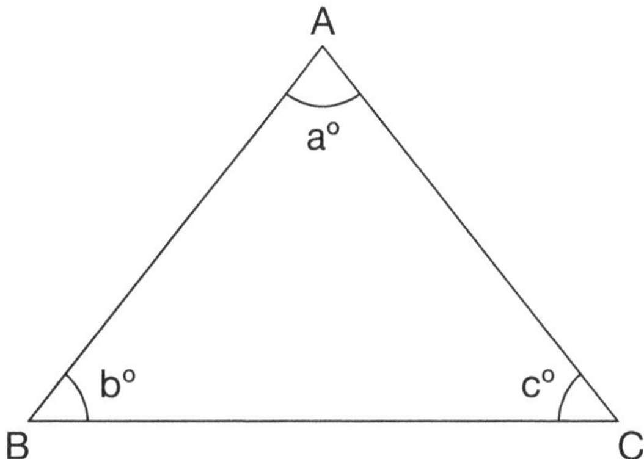

✓ Interior angles; $\angle a^\circ$, $\angle b^\circ$, $\angle c^\circ$,

✓ $m \angle a^\circ + m \angle b^\circ + m \angle c^\circ = 180^\circ$

Exterior angles: The exterior angle is equal to the sum of the two opposite interior angles.

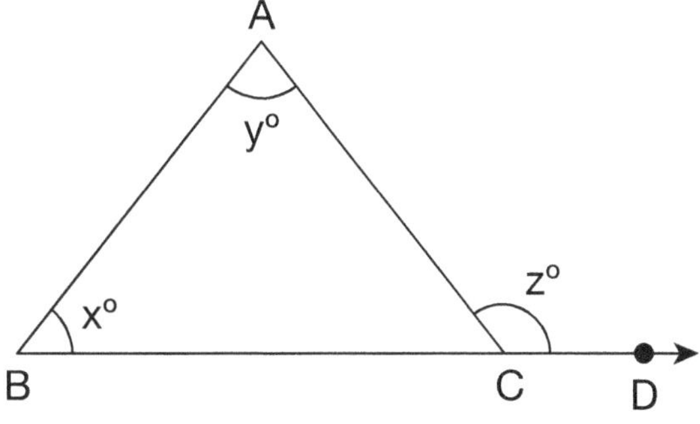

✓ In the figure $\angle ACD$ is an exterior angle of $\triangle ABC$.

✓ $m \angle z = m \angle x + m \angle y$

INTERIOR AND EXTERIOR TRIANGLES TEST

1. In $\triangle$ ABC, m $\angle$ B = 35° and m $\angle$ A = 53°
Find $\angle$ x.

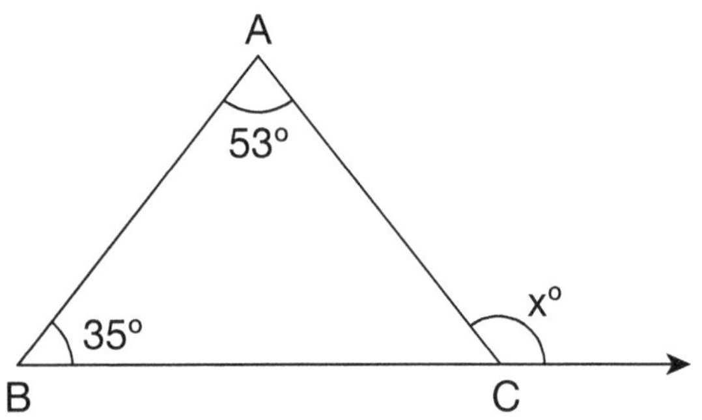

A) 55°

B) 75°

C) 88°

D) 92°

2. Find the value of x in the figure below.

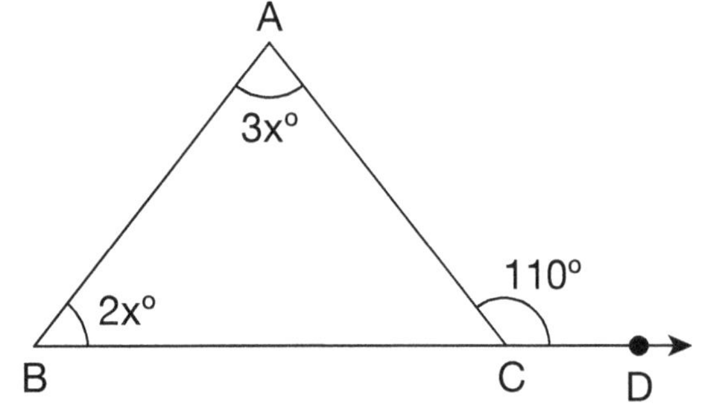

A) 22°

B) 32°

C) 42°

D) 48°

3. Find the value of x in the figure below.

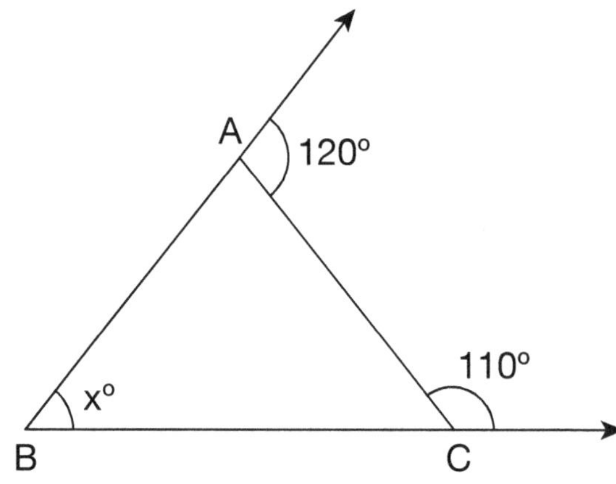

A) 40°

B) 50°

C) 80°

D) 110°

4. Find the value of x in the figure below.

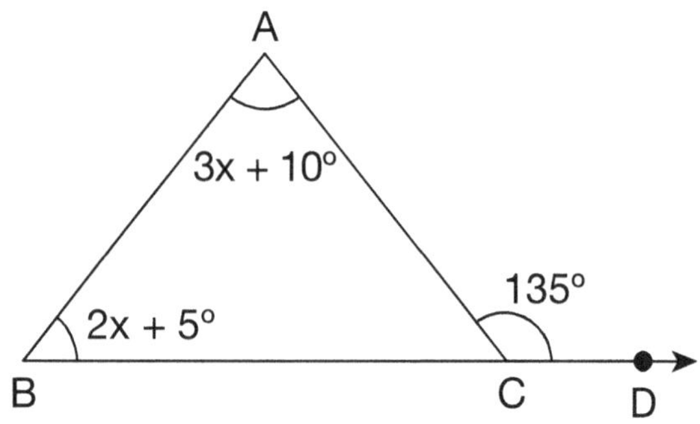

A) 10°

B) 14°

C) 22°

D) 24°

5. Find the value of x in the figure below.

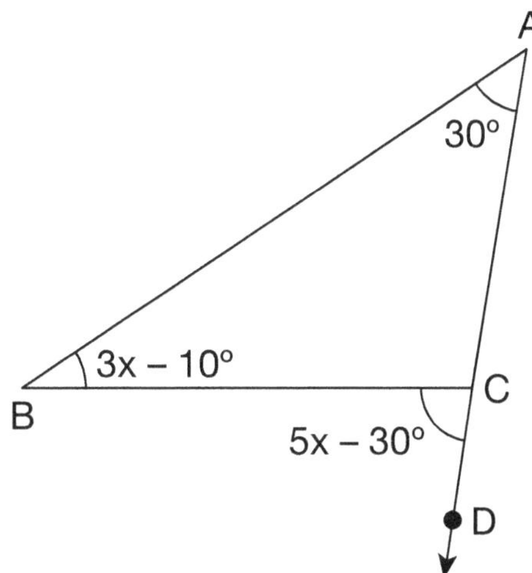

A) 15°

B) 25°

C) 35°

D) 45°

TRIANGLE INEQUALITIES

Triangle inequality:

The sum of the lengths of any two sides of a triangle is greater than the length of the third side.

The difference of the lengths of any two sides of a triangle is less than the length of the third side.

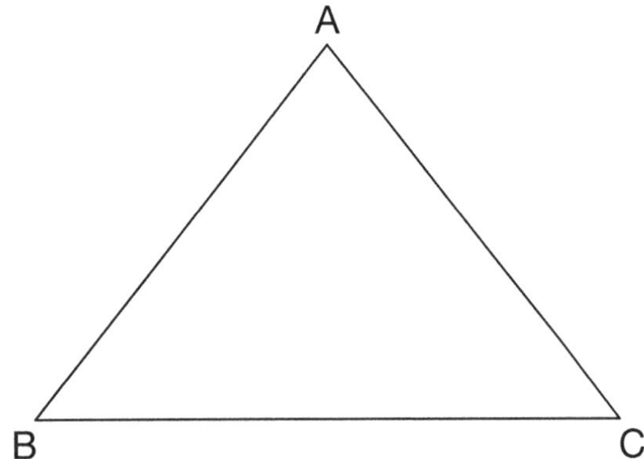

$|b - c| < a < b + c$

$|a - c| < b < a + c$

$|a - b| < c < a + b$

Example: Which of the following can be x?

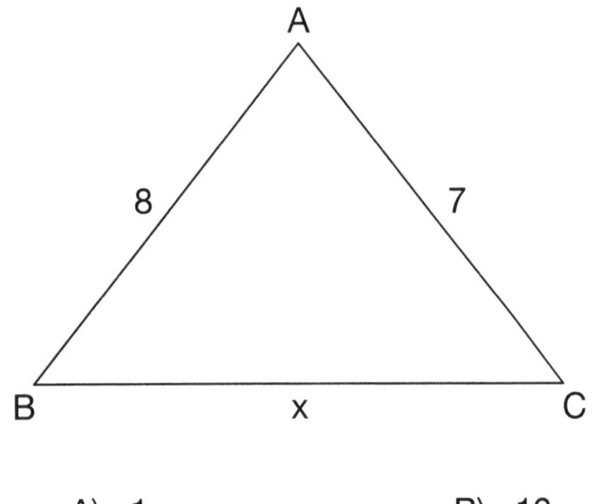

A) 1　　B) 12　　C) 15　　D) 16

Solution:

From triangle inequality theorem:

$8 - 7 < x < 8 + 7$

$1 < x < 15$, x only can be 12 from all options.

TRIANGLE INEQUALITIES TEST

1. Which of following is the smallest angle in this triangle?

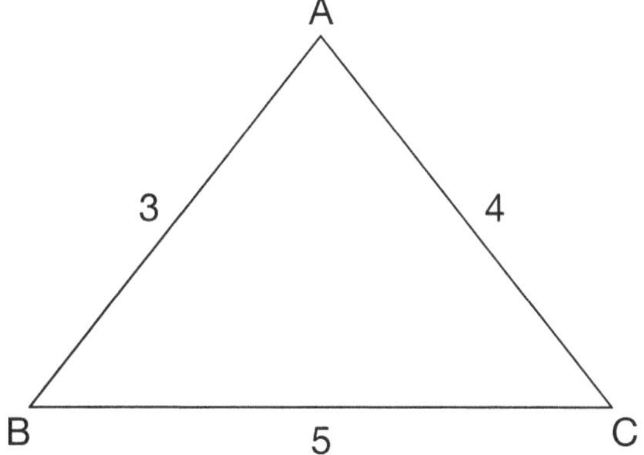

A) < A

B) < B

C) < C

D) none

2. Which of the following is the sides of △ ABC from shortest to longest?

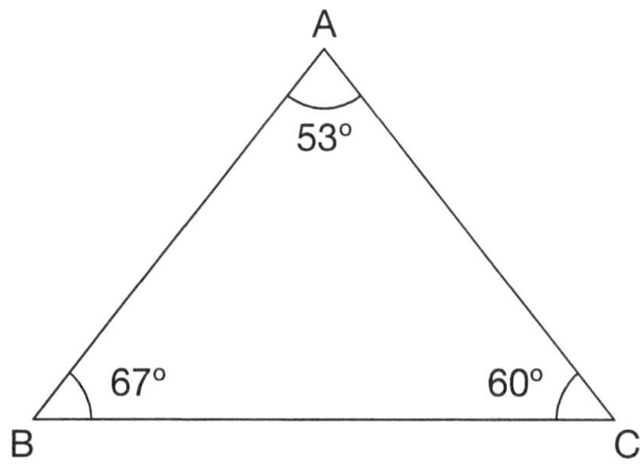

A) $\overline{AB} < \overline{BC} < \overline{AC}$

B) $\overline{AC} < \overline{BC} < \overline{AB}$

C) $\overline{BC} < \overline{AB} < \overline{AC}$

D) $\overline{AB} < \overline{AC} < \overline{BC}$

3. A triangle has two sides with a length of 12 and 15. Which of following is the range of possible values for the third side?

A) $1 < x < 12$

B) $3 < x < 15$

C) $3 < x < 25$

D) $3 < x < 27$

4. Which side has the greatest measure ıf $\angle z = 110°$ and $\angle y = \angle z - 50°$?

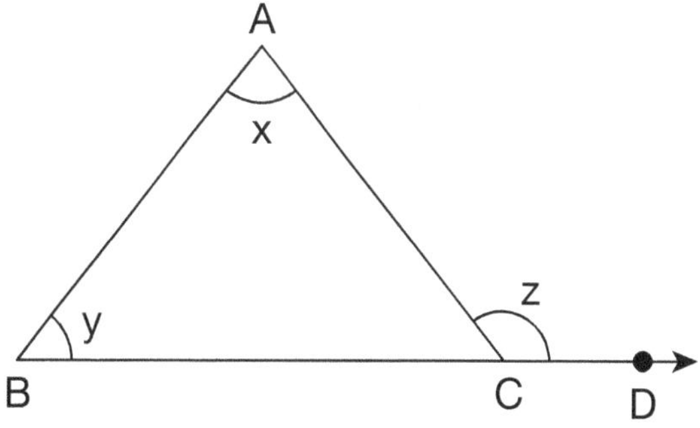

A) $\overline{AB}$

B) $\overline{BC}$

C) $\overline{AC}$

D) $\overline{CD}$

5. Which of following is the one possible value of x?

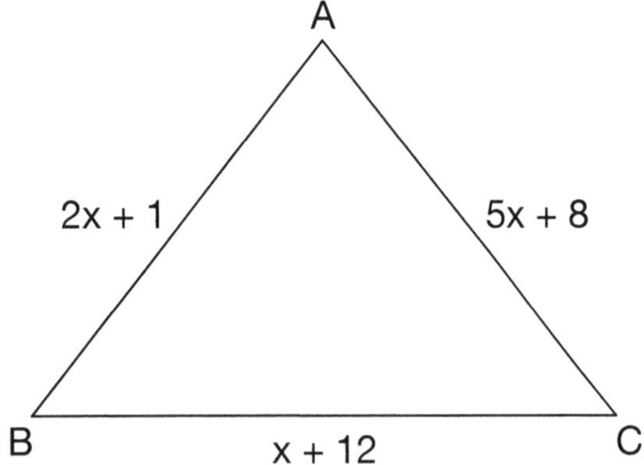

A) –3

B) 0

C) 2

D) 3

SPECIAL RIGHT TRIANGLES

30°–60°–90° Special right triangle: In a triangle 30°–60°– 90°, the hypotenuse is twice as long side as short side, and the longer side is $\sqrt{3}$ times as long as the shorter side.

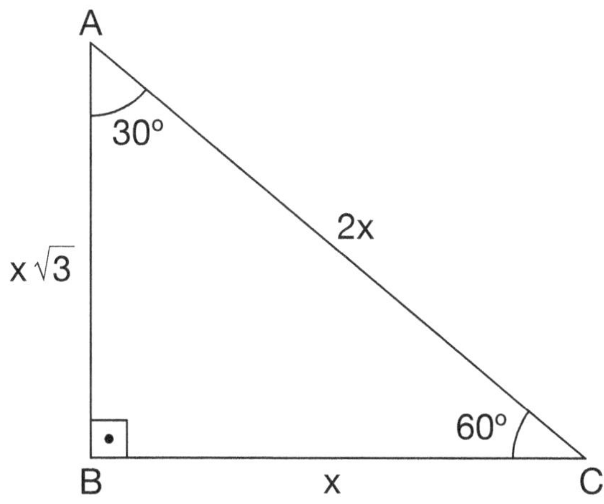

45°–45°–90° Special right triangle: In a triangle 45°–45°–90°, the hypotenuse is $\sqrt{2}$ times the length of either of the other two sides, which are both equal.

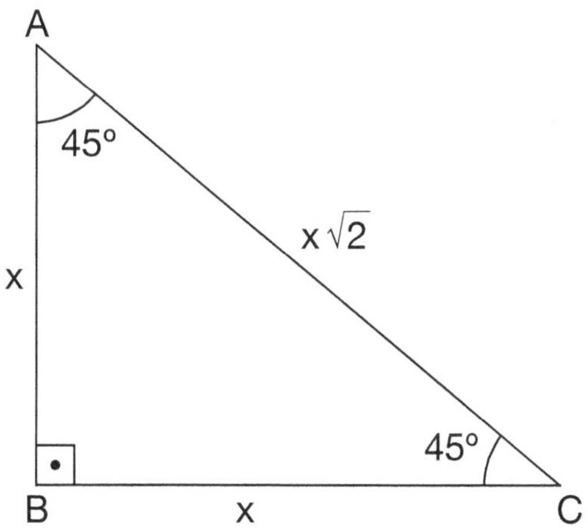

SPECIAL RIGHT TRIANGLES TEST

1. Find the value of x.

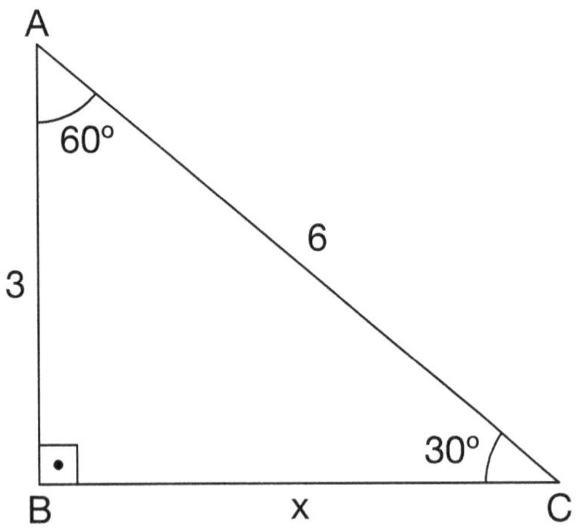

A) 3

B) $\sqrt{3}$

C) $3\sqrt{3}$

D) 6

2. Find the value of x.

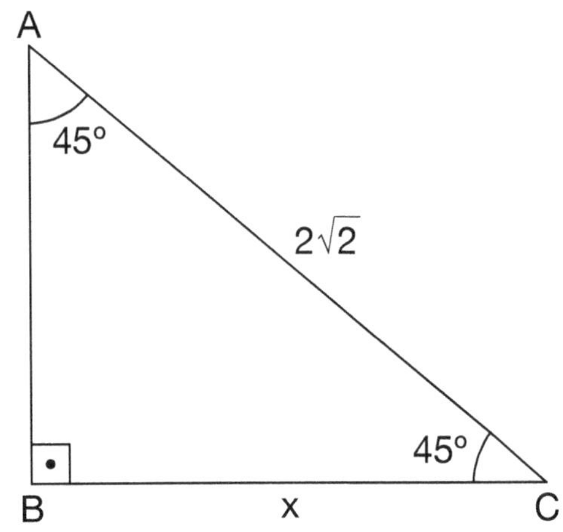

A) 2

B) $\sqrt{2}$

C) $2\sqrt{2}$

D) 4

3. Find the value of x.

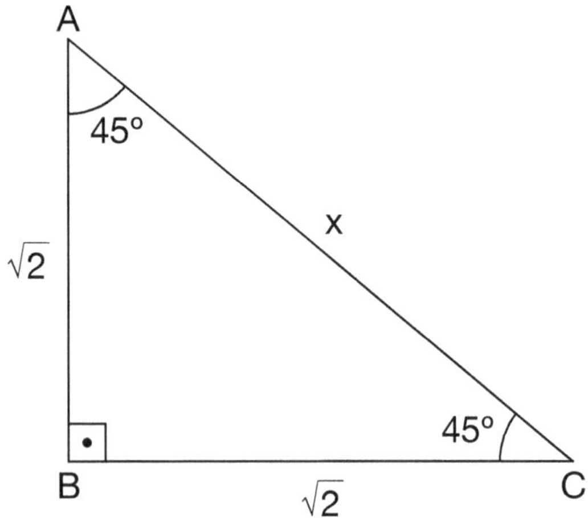

A) 1

B) 2

C) 3

D) 4

4. Find the value of x.

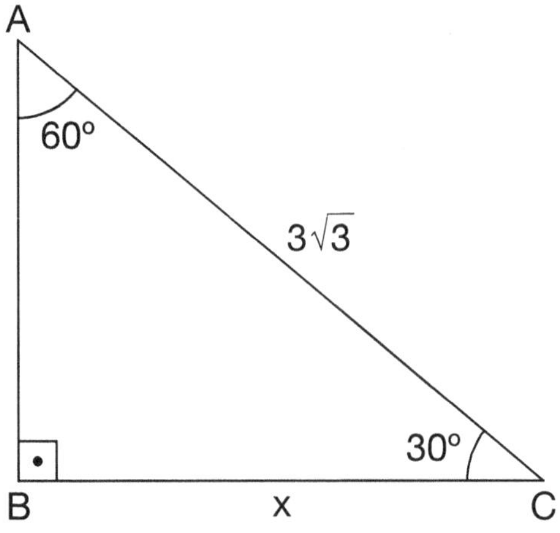

A) $\frac{1}{2}$

B) $\frac{3}{2}$

C) $\frac{7}{2}$

D) $\frac{9}{2}$

SPECIAL RIGHT TRIANGLES TEST

5. Find the value of x.

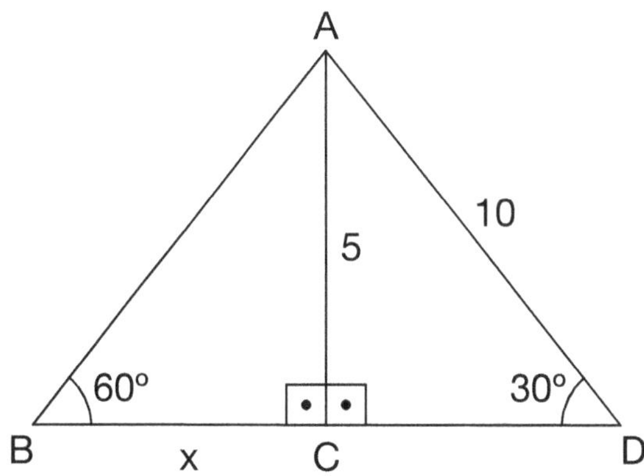

A) 5

B) $5\sqrt{3}$

C) $\frac{5\sqrt{3}}{3}$

D) 10

American Math Academy

PYTHAGOREAN THEOREM

In a right triangle, the square of the hypotenuse (long side) is equal to the sum of the squares of the other two sides.

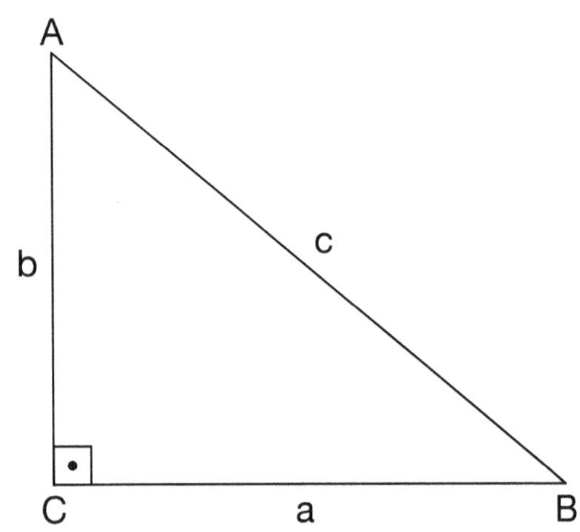

$$a^2 + b^2 = c^2$$

Example: What is the value of x?

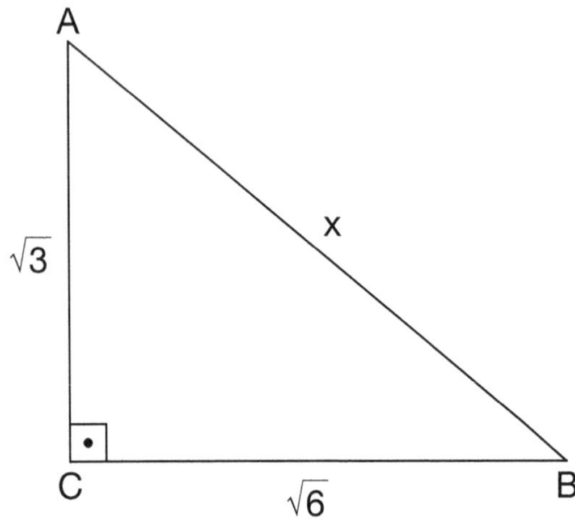

Solution:

$a^2 + b^2 = c^2$

$(\sqrt{3})^2 + (\sqrt{6})^2 = x^2$

$3 + 6 = x^2$

$9 = x^2$

$3 = x$

PYTHAGOREAN THEOREM TEST

1. Find the hypotenuse of a triangle with a base of 12in and height of 9in.

A) 5 in

B) 12 in

C) 13 in

D) 15 in

2. American Math Academy Center is 16m long and 12m wide. How long is the diagonal of the American Math Academy?

A) 12m

B) 16m

C) 15m

D) 20m

American Math Academy

3. What is the value of x?

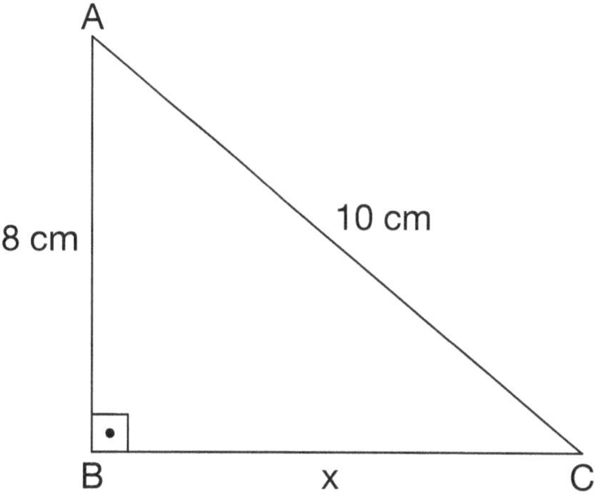

A) 6 cm

B) 8 cm

C) 10 cm

D) 15 cm

4. What is the value of x + y?

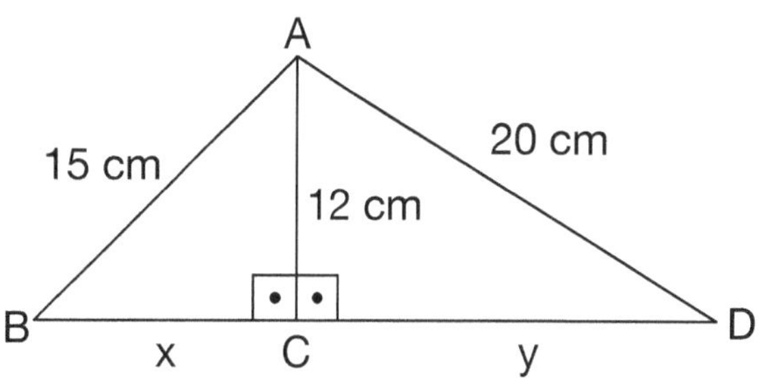

A) 12 cm

B) 15 cm

C) 20 cm

D) 25 cm

5. What is the value of x?

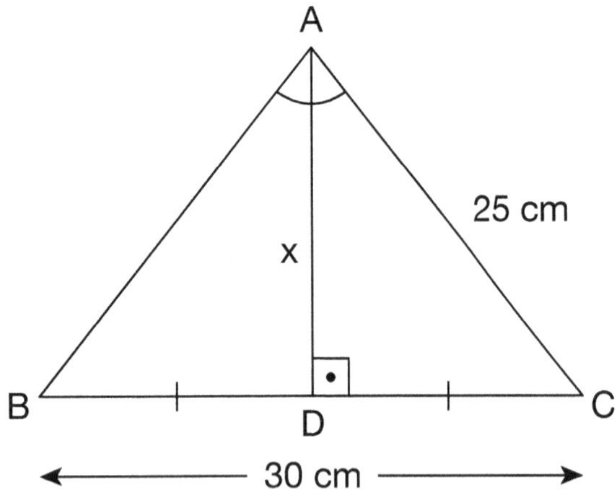

A) 12 cm

B) 15 cm

C) 16 cm

D) 20 cm

MIDPOINT AND DISTANCE

MIDPOINT:

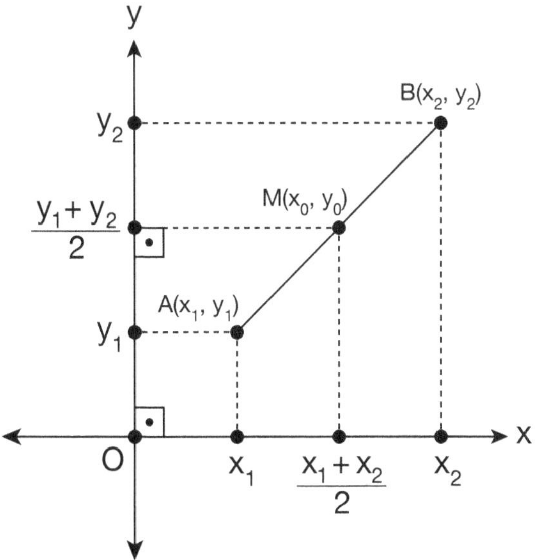

What are coordinates (X_0, Y_0) of the midpoint of the segment whose endpoints are $A(x_1, y_1)$ and $B(x_2, y_2)$?

$$x_0 = \frac{x_1+x_2}{2}$$

$$y_0 = \frac{y_1+y_2}{2}$$

DISTANCE:

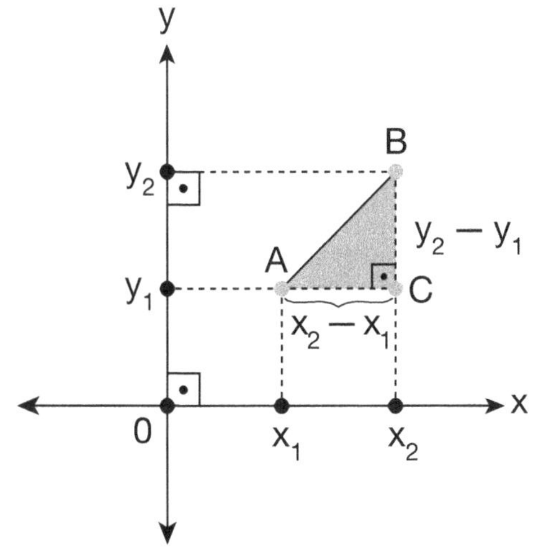

Use the pythagorean theorem to find the length of $\overline{AB}$, which is the hypotenuse of the right triangle.

$$|AB|^2 = (x_2 - x_1)^2 + (y_2 - y_1)^2$$

$$|AB| = \sqrt{(x_2 - x_1)^2 + (y_2 - y_1)^2}$$

MIDPOINT AND DISTANCE TEST

1. Find the distance between the following two points:

(9, 7) and (7, 3)

A) 5

B) $\sqrt{5}$

C) $2\sqrt{5}$

D) 10

2. Calculate the midpoint of the segment with the given endpoints:

(9, 7) and (7, 3)

A) (8, 3)

B) (8, 5)

C) (9, 7)

D) (5, 8)

3. The midpoint between the points A (4, 8) and B (a, b) is (12, 16).

What are the coordinates of point B?

A) (20, 24)

B) (24, 20)

C) (–24, –20)

D) (–20, –24)

American Math Academy

4. On the coordinate plane below, what is the midpoint of AB?

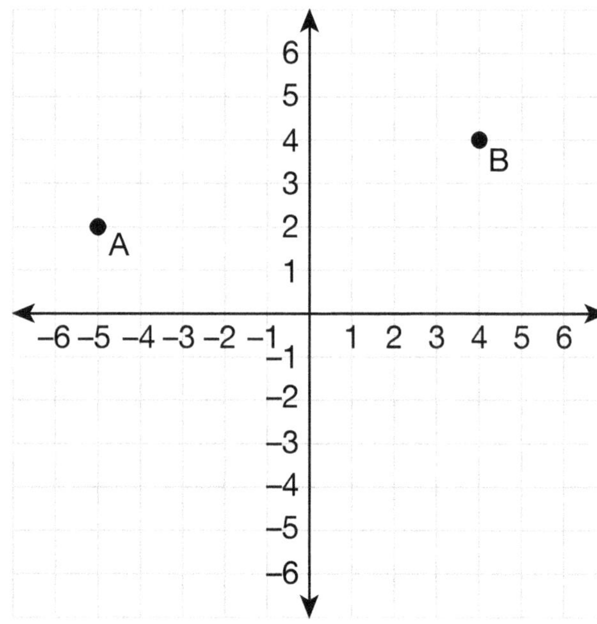

A) (0.5, –3)

B) (0.5, 3)

C) (–0.5, 3)

D) (–0.5, –3)

5. Find the midpoint of (–1, –13) and (9, 23).

A) (4, 5)

B) (4, –5)

C) (–4, –5)

D) (–4, 5)

COORDINATE PLANE

Coordinate Plane: A coordinate system formed by the intersection of a vertical number line, called the y–axis, and a horizontal number line, called the x–axis.

Origin: A beginning or starting point and lines that intersect each other at zero (0,0).

Quadrant: One of four regions into which the coordinate plane is divided by the x– and y–axis. These regions are called quadrants.

Ordered pair: A pair of numbers that can be used to locate a point on a coordinate plane. The order of the numbers in a pair is important, and the x–axis always comes before the y–axis.

x–coordinate: The first number in an ordered pair is called the x–coordinate, and it is the first value in an ordered pair.

y–coordinate: The second number in an ordered pair is called the y–coordinate. It is the second value in an ordered pair.

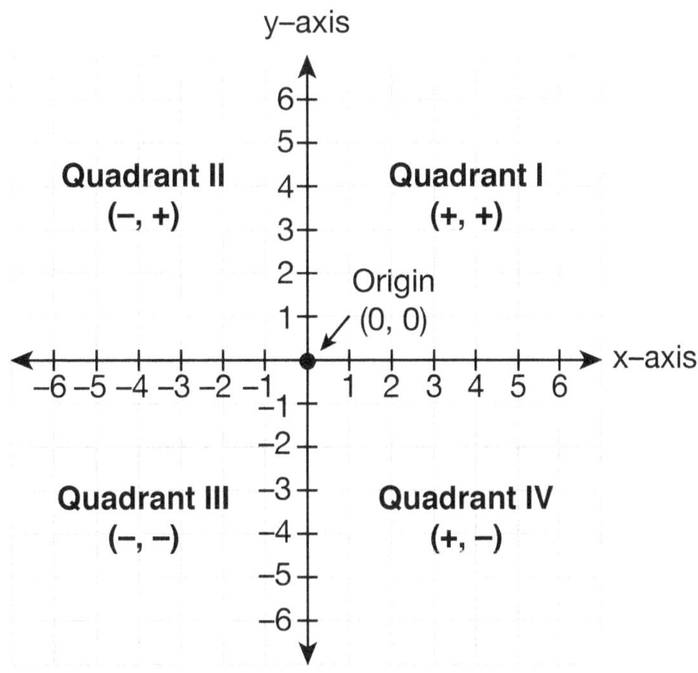

COORDINATE PLANE TEST

1. Name the quadrant that point (–3,4) is in.

A) Quadrant I

B) Quadrant II

C) Quadrant III

D) Quadrant IV

2. Which ordered pair locates a point on the y–axis?

A) (2, 2)

B) (1, 0)

C) (4, 0)

D) (0, –2)

3. The distance from (9, 12) to the origin is

A) 9 units

B) 12 units

C) 13 units

D) 15 units

4. What are the coordinates of point C?

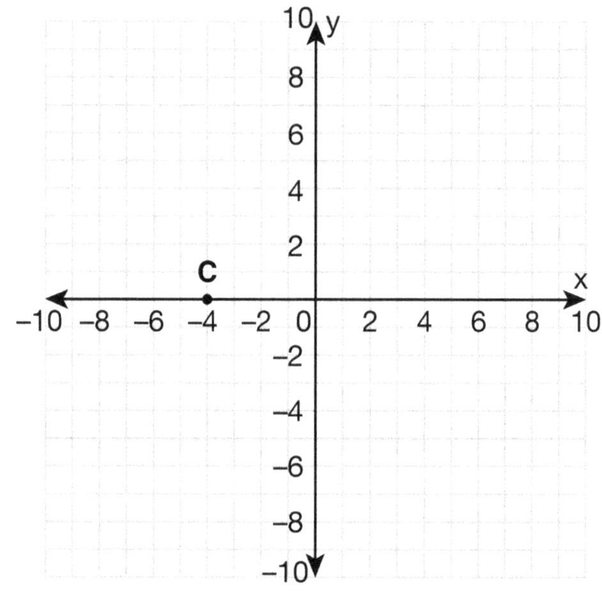

A) (0, 4)

B) (0, 0)

C) (0, –4)

D) (–4, 0)

5. Name the quadrant that point (6, 0) is in.

A) Quadrant I

B) Quadrant II

C) Quadrant III

D) No Quadrant

SLOPE AND SLOPE INTERCEPT FORM

Slope Intercept Form:

$y = mx + b$ m : slope

b : y–intercept = (0, b)

Suppose there are two points on a line, (x_1, y_1) and (x_2, y_2).

The slope m of the line is:

$$m = \frac{\text{change in y (Rise)}}{\text{change in x (Run)}} = \frac{y_2 - y_1}{x_2 - x_1}$$

Example:

Find the equation of the line that passes through the points (1, 2) and (3, 6)

Solution:

Slope = m = $\frac{6-2}{3-1} = \frac{4}{2} = 2$

Example: Find the slope of $y = 3x - 7$

Solution: If $y = mx + b$, then $m = 3$

Example: Find the slope of $y = \frac{1}{7}x + 13$

Solution: The equation is in slope intercept form ($y = mx + b$), so slope = m = $\frac{1}{7}$

SLOPE AND SLOPE INTERCEPT FORM TEST

1. Which of the following equations represents a line that is perpendicular to the line that passes through points A(3, 6) and B(4, 8)?

A) $y = 2x - 5$

B) $y = -\frac{1}{2}x + 5$

C) $y = 2x + 5$

D) $y = -x - 5$

2. Which of the following is the slope of the graph?

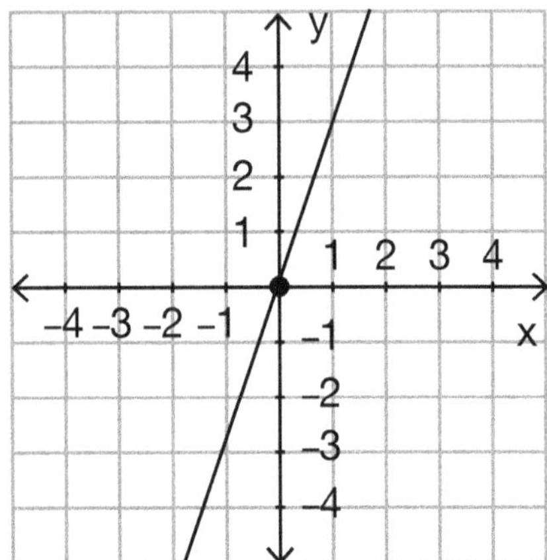

A) 3

B) –3

C) $\frac{1}{3}$

D) 0

3. Line M passes through the coordinate points $\left(-2, \frac{1}{3}\right)$ and $\left(1, \frac{-2}{6}\right)$. What is the slope of line M?

A) $\frac{1}{9}$

B) $-\frac{2}{9}$

C) $-\frac{9}{2}$

D) –9

4. What is the value of k if a line passing through (6, 3) and (2,k) has a slope of –1?

A) 3

B) 5

C) 7

D) 9

5. What is the slope of the line that passes through (3, 2) and (1, 8)?

A) 2

B) 12

C) 3

D) –3

6. Which of the following equations is a line that passes through the coordinates (0, 3) and (3, 2)?

A) $y = -2x$

B) $y = 3x - 1$

C) $y = -\frac{1}{3}x + 3$

D) $y = \frac{1}{3}x + 3$

SIMILARITY THEOREM

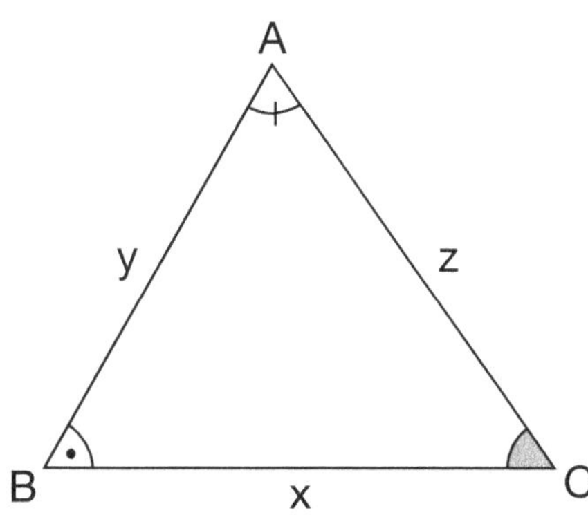

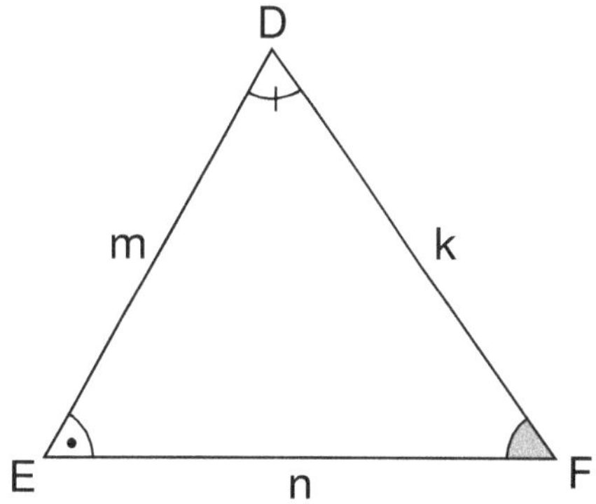

When two figures are similar, the ratios of the lengths of their corresponding sides are equal.

If $\widehat{ABC} \sim \widehat{DEF}$ then $\angle A = \angle D$ and $\frac{|AB|}{|DE|} = \frac{|AC|}{|DF|} = \frac{|BC|}{|EF|}$

$\angle B = \angle E$

$\angle C = \angle F$

$$\frac{x}{n} = \frac{y}{m} = \frac{z}{k}$$

Note:

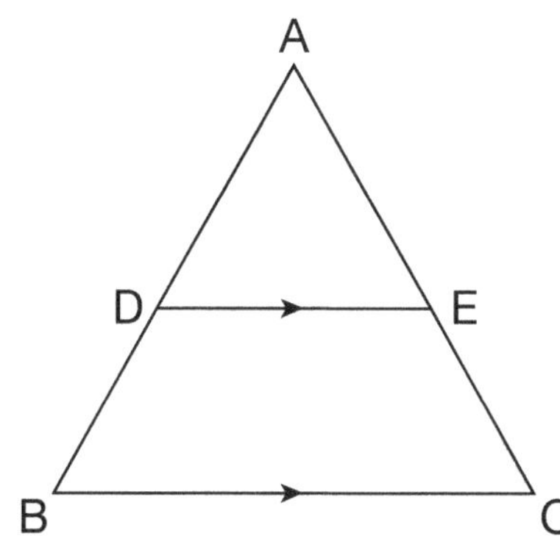

If [DE] // [BC], then

$$\frac{|AD|}{|AB|} = \frac{|DE|}{|BC|} = \frac{|AE|}{|AC|}$$

SIMILARITY THEOREM TEST

1.

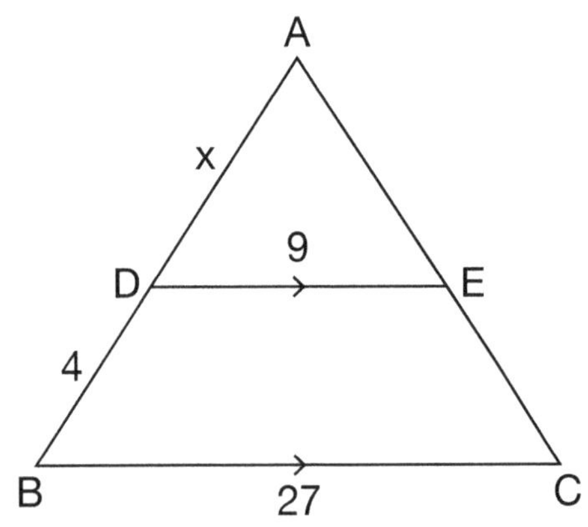

If DE // BC in the figure above, what is the value of x?

A) 2

B) 4

C) 6

D) 8

2.

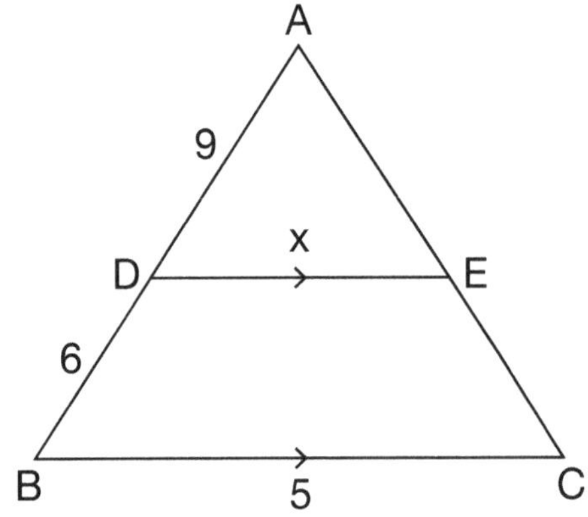

If DE // BC in the figure above, what is the value of x?

A) 3

B) 4

C) 5

D) 6

American Math Academy

3.

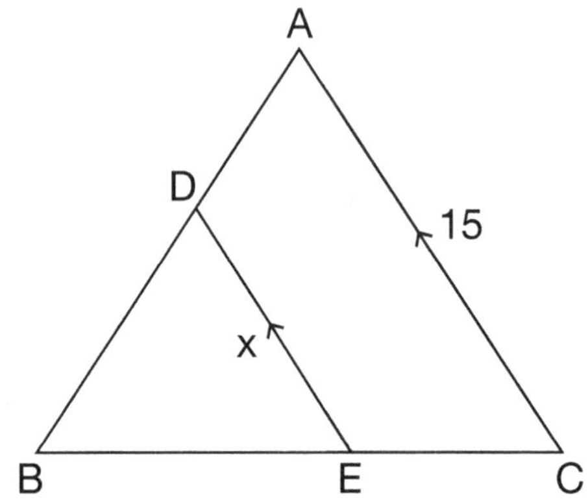

If DE // AC and 3|AD| = 2|BD| in the figure above, what is the value of x?

A) 3

B) 6

C) 9

D) 12

4.

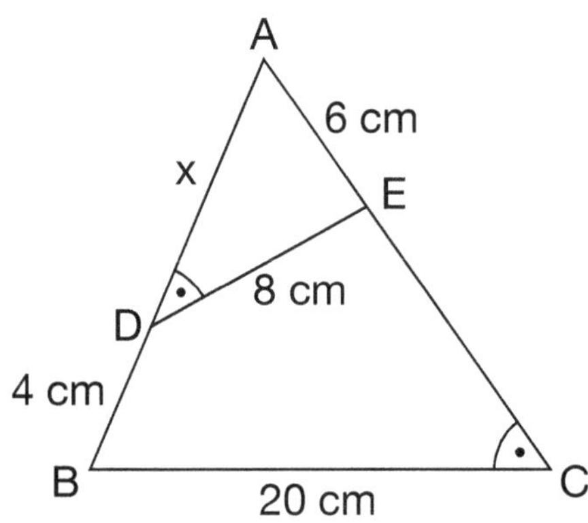

From the figure above, if ∠ ADE ≅ ∠ ACB, and |BC| = 20cm, |DE| = 8cm, |EA| = 6cm, and |BD| = 4cm, then what is the value of x?

A) 5 cm

B) 9 cm

C) 11 cm

D) 15 cm

5.

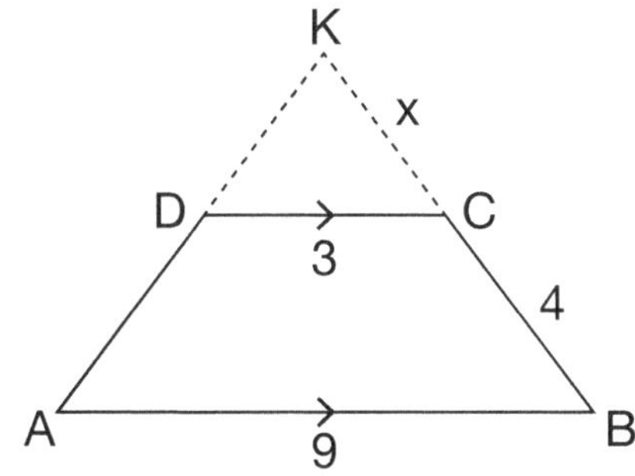

If DC // AB in the figure above, what is the value of x?

A) 2

B) 4

C) 6

D) 8

6.

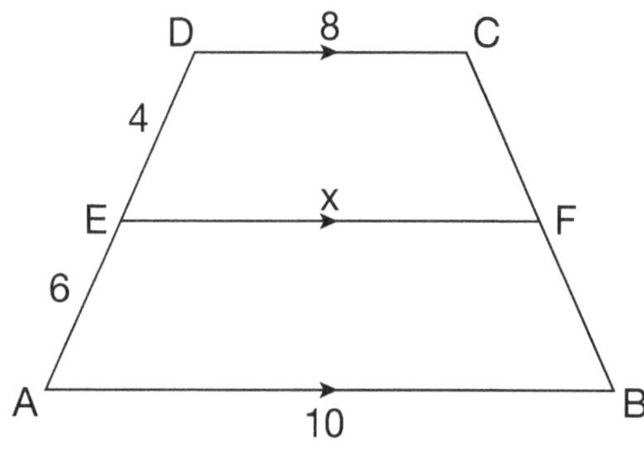

If DC // EF // AB in the trapezoid ABCD above, DC = 8, EF = x, and AB = 10.
what is the value of x?

A) 2.2

B) 4.4

C) 8.8

D) 9.8

AREA AND PERIMETER OF TRIANGLE

Area of right triangle:

$$\text{Area} = \frac{a \times b}{2}$$

Area of an equilateral triangle:

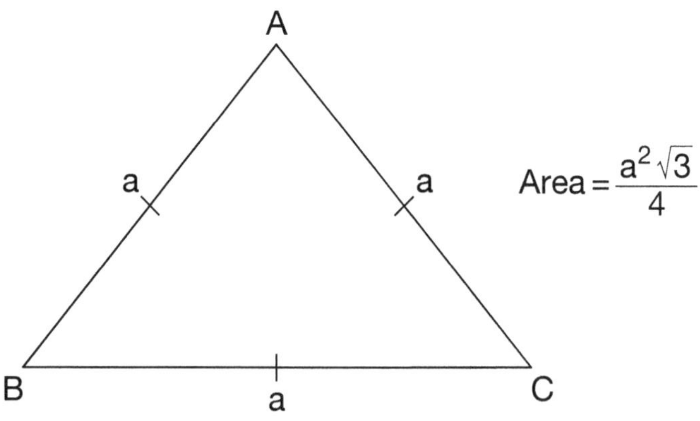

$$\text{Area} = \frac{a^2\sqrt{3}}{4}$$

Are of isosceles triangle:

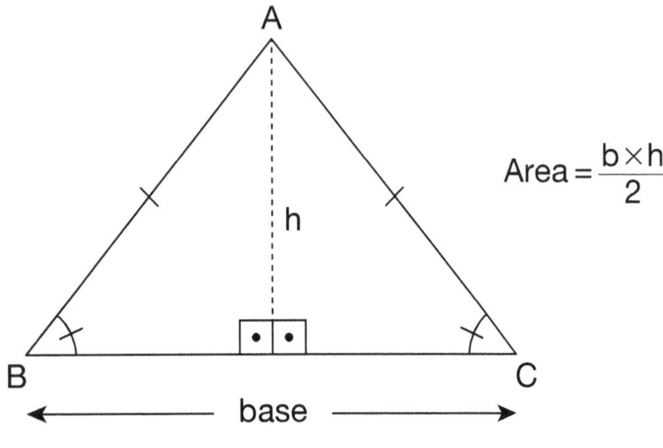

$$\text{Area} = \frac{b \times h}{2}$$

Heron's formula:

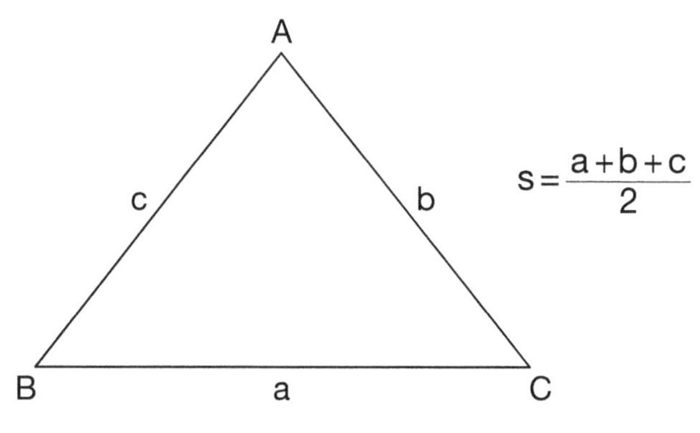

$$s = \frac{a+b+c}{2}$$

$$\text{Area} = \sqrt{s(s-a)\cdot(s-b)\cdot(s-c)}$$

AREA AND PERIMETER OF TRIANGLE

Perimeter of scalene triangle:

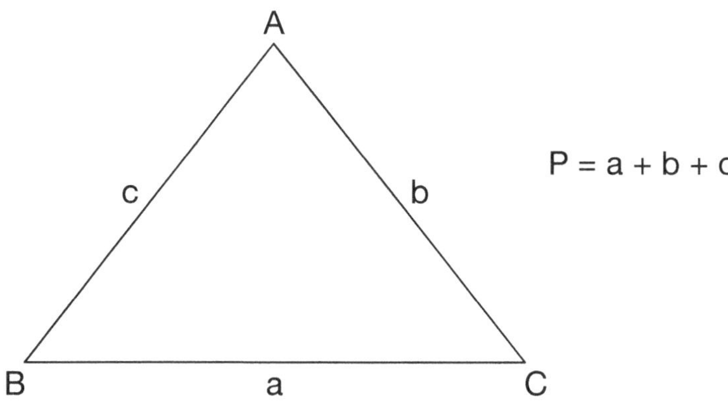

$P = a + b + c$

Perimeter of isosceles triangle:

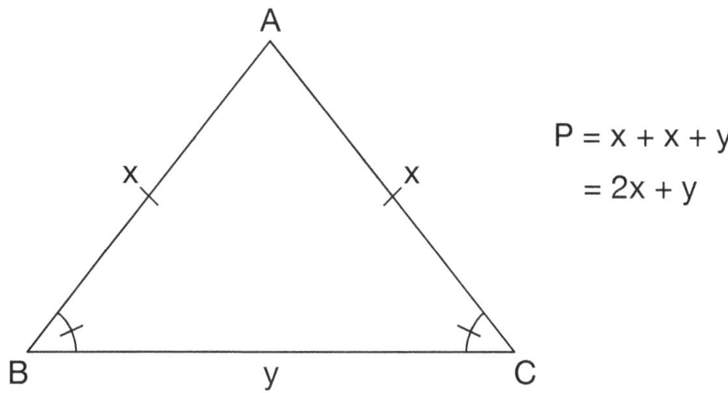

$P = x + x + y$
$= 2x + y$

Perimeter of equilateral triangle:

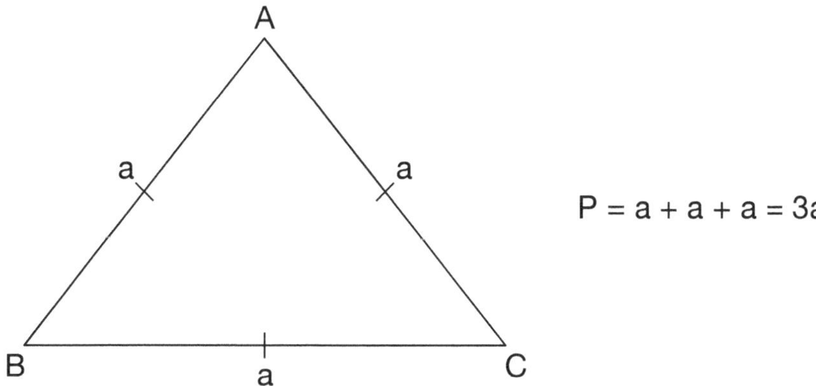

$P = a + a + a = 3a$

AREA AND PERIMETER OF TRIANGLE TEST

1. What is the area of $\triangle ABC$?

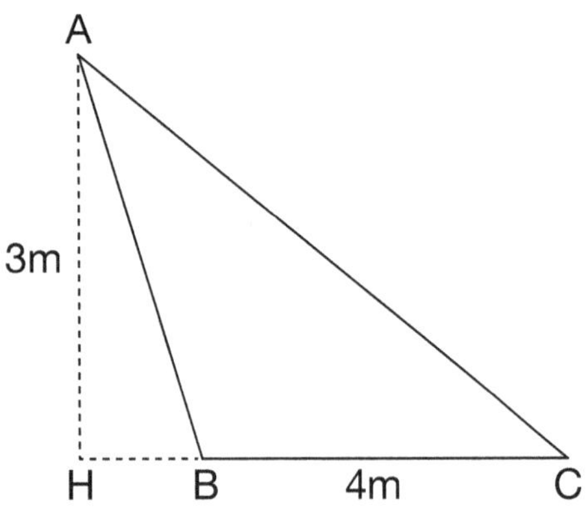

A) $6m^2$

B) $12m^2$

C) $18m^2$

D) $24m^2$

2. Find the area of $\triangle ABC$?

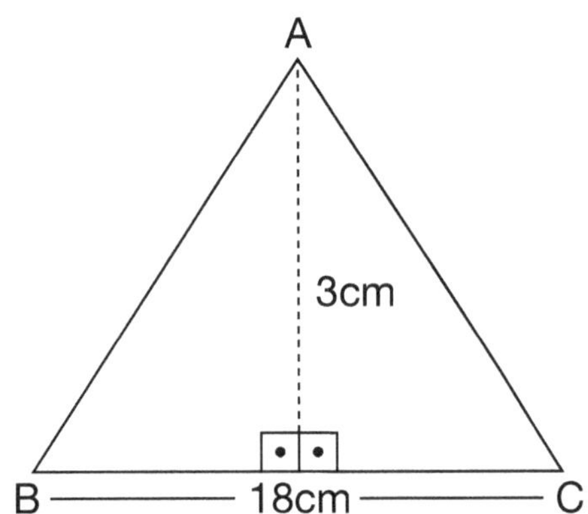

A) $18cm^2$

B) $27cm^2$

C) $30cm^2$

D) $36cm^2$

3. What is the area of an equilateral triangle with a side of 4?

A) $\sqrt{3}$

B) $2\sqrt{3}$

C) $4\sqrt{3}$

D) $6\sqrt{3}$

4. Find the perimeter of the following triangle

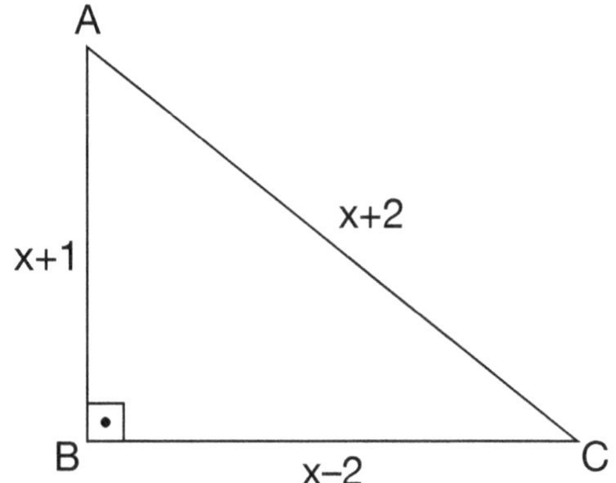

A) $3x$

B) $3x - 1$

C) $3x + 1$

D) $3x + 2$

AREA AND PERIMETER OF TRIANGLE TEST

5. Find the area of the triangle (Use Heron's Formula)

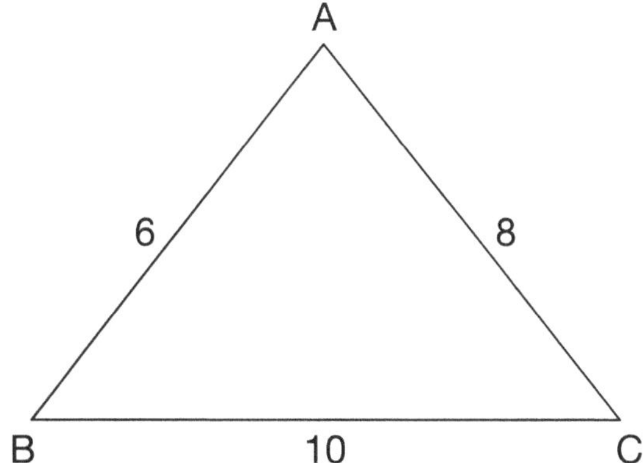

A) $\sqrt{15}$

B) $3\sqrt{15}$

C) 9

D) 24

6. Find the area of the triangle with a base of 8 in and perpendicular height of 26 in?

A) 52 in^2

B) 104 in^2

C) 208 in^2

D) 260 in^2

7. What is the area of Δ ABC?

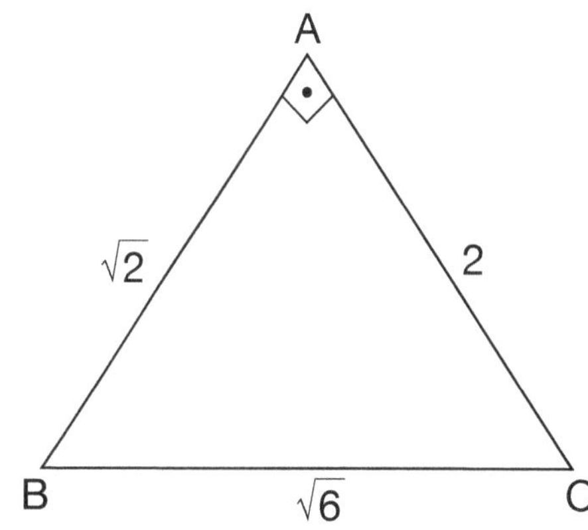

A) 2

B) $\sqrt{2}$

C) $2\sqrt{2}$

D) $3\sqrt{2}$

AREA AND PERIMETER OF QUADRILATERALS

Parallelogram:

Area = base × height

P = 2(h + b)

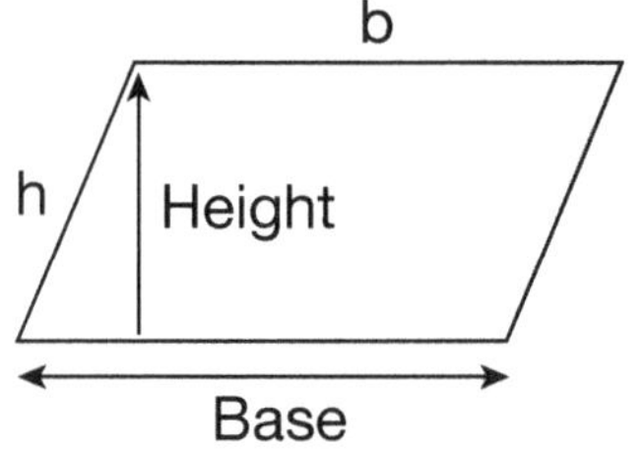

Rectangle:

Area = base × height = b × h

P = 2(h + b)

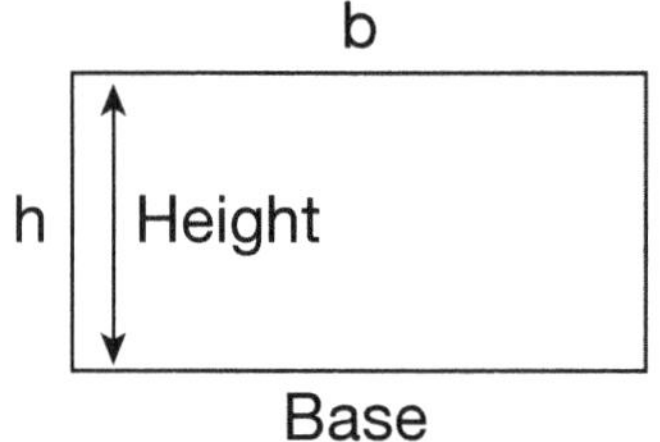

Square:

Area = width × length = s × s

P = 4s

s

s Area = s^2

Trapezoid:

$$\text{Area} = \frac{(b_1 + b_2)}{2} \times h$$

$$P = m + n + b_1 + b_2$$

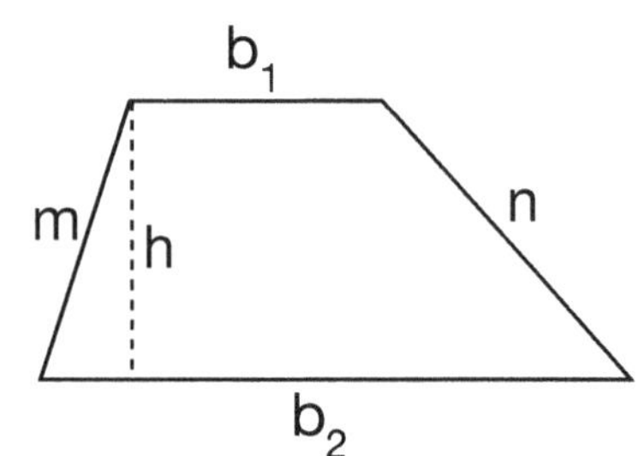

AREA AND PERIMETER OF QUADRILATERALS TEST

1. 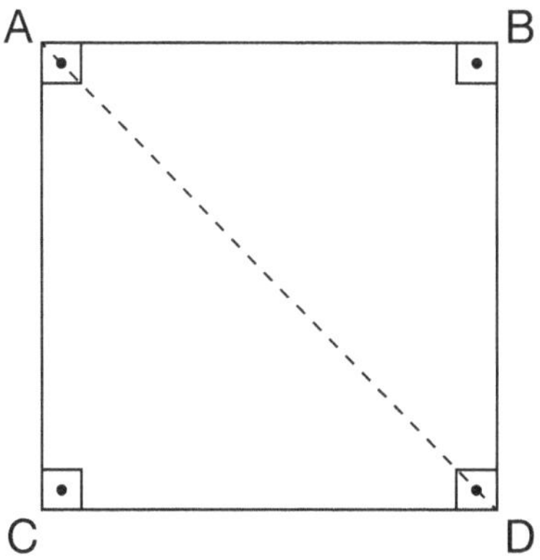

What is the area of the above square if the length of AD is $6\sqrt{2}$?

A) 16

B) 25

C) 36

D) 47

2. 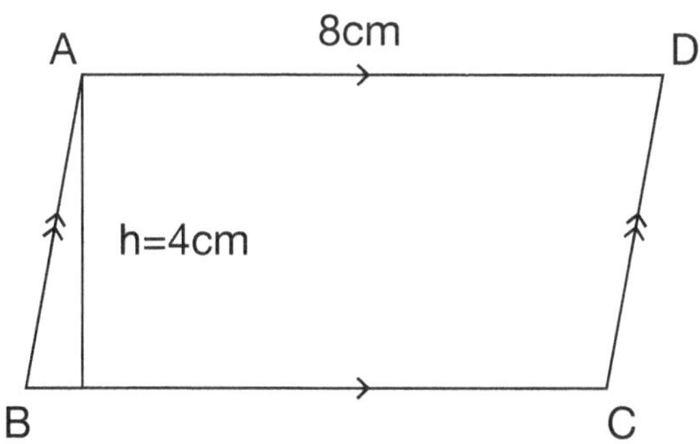

What is the area of parallelogram ABCD?

A) 32 cm^2

B) 48 cm^2

C) 52 cm^2

D) 64 cm^2

3. The perimeter of a rectangle is 48m. If the width of the rectangle is three times the length, what is the width?

A) 6 cm

B) 9 cm

C) 15 cm

D) 18 cm

4. 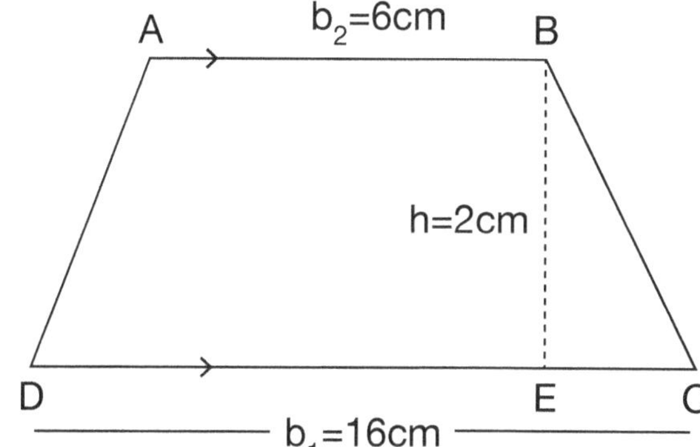

What is the area of trapezoid ABCD?

A) 11 cm^2

B) 22 cm^2

C) 33 cm^2

D) 42 cm^2

5.

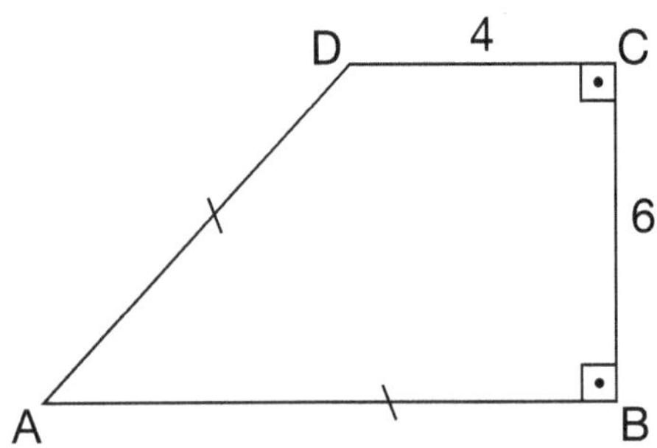

In the above trapezoid ABCD,
DC = 4, BC = 6, and AD = AB.
What is the value of |AB|?

A) $\frac{13}{4}$

B) $\frac{13}{2}$

C) 13

D) 41

6.

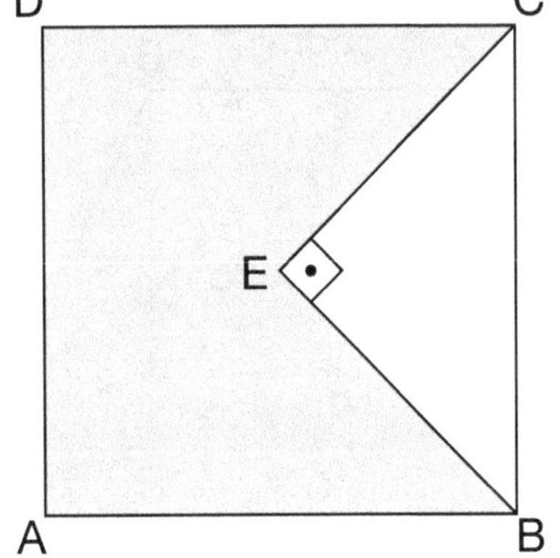

Square ABCD has a perimeter of 32m, and the perimeter of the triangle EBC is 18m. Find the area of the shaded part.

A) 18 m^2

B) 52 m^2

C) 55 m^2

D) 64 m^2

7. What is the area of ABCD?

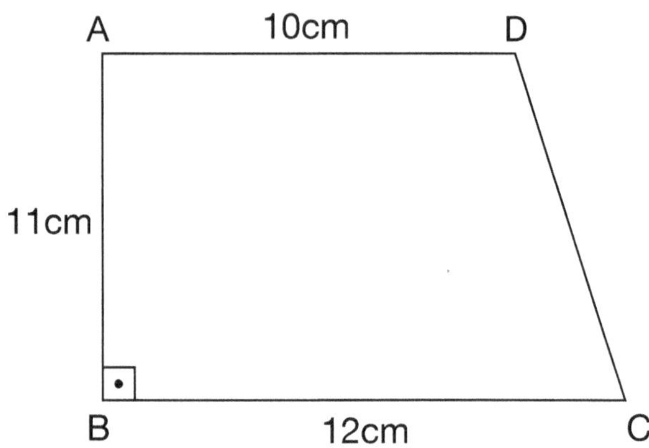

A) 22cm^2

B) 44cm^2

C) 108cm^2

D) 121cm^2

American Math Academy

8.

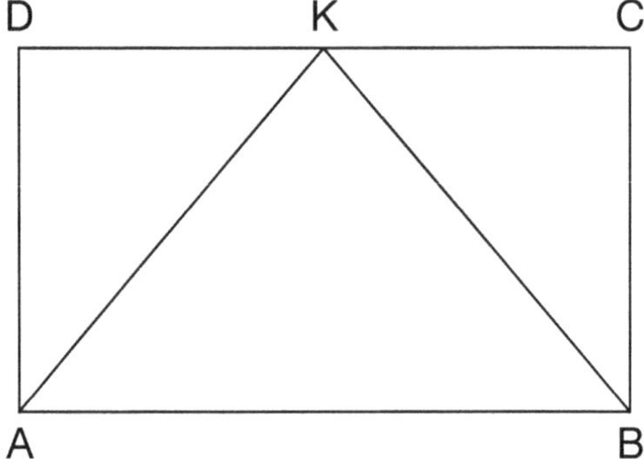

The figure ABCD is a rectangle and AKB is an equilateral triangle. If the area of the rectangle is $8\sqrt{3}$, then find the length of AB.

A) 2

B) 4

C) 8

D) 12

TRANSFORMATIONS

Translation: A translation is a change in location with the same distance and same direction.

Example: In the example below triangle X is translated to become triangle Y

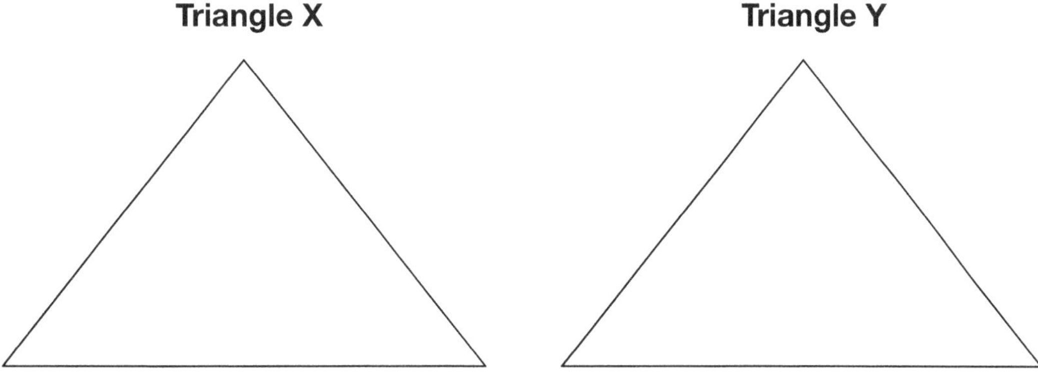

Rotation: A rotation means turn a figure.

Example: Triangle X was rotated right 90° to make triangle Y.

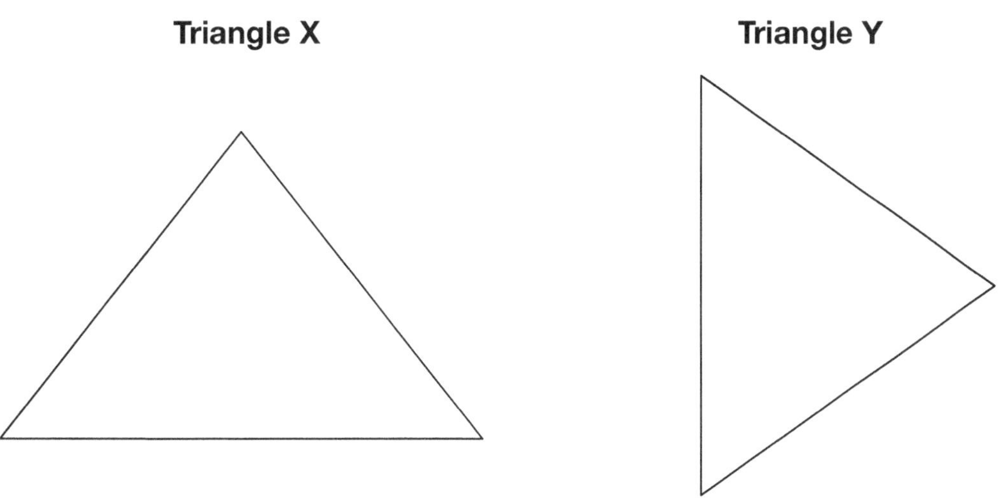

Reflection: A reflection is a transformation that flips a figure across a line.

Example: A reflection is flipped over a line.

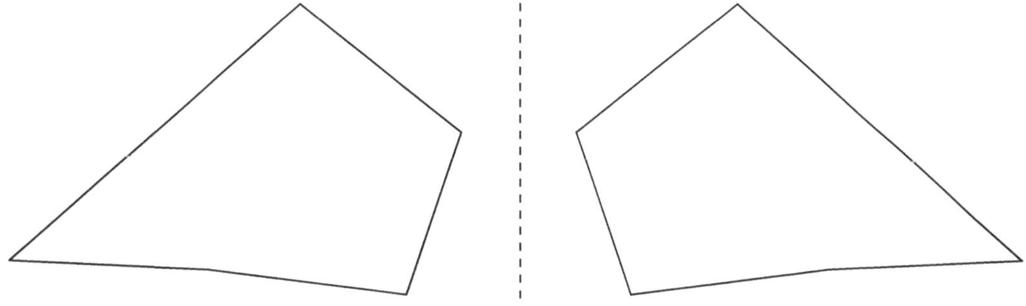

Type of reflection	Point before reflection	Point after reflection
Across x–axis	**(x, y)**	**(x , – y)**
Across y–axis	**(x, y)**	**(– x , y)**
Across the line y=x	**(x , y)**	**(y , x)**
Across the line y=–x	**(x , y)**	**(– y, –x)**
In origin	**(x , y)**	**(– x , –y)**

Dilation: Dilation is when the size of the shape changes without any change to the shape itself.

Example:

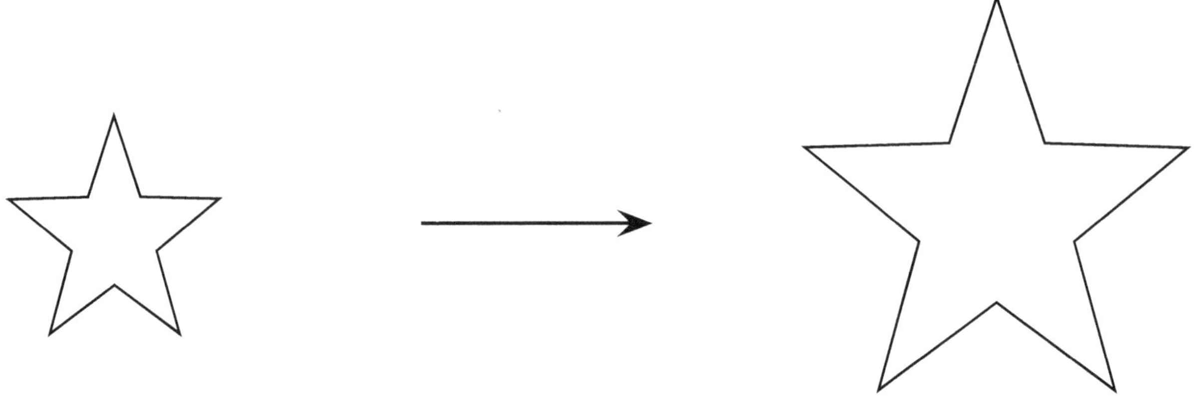

TRANSFORMATIONS TEST

1. Point M is plotted in the graph below.

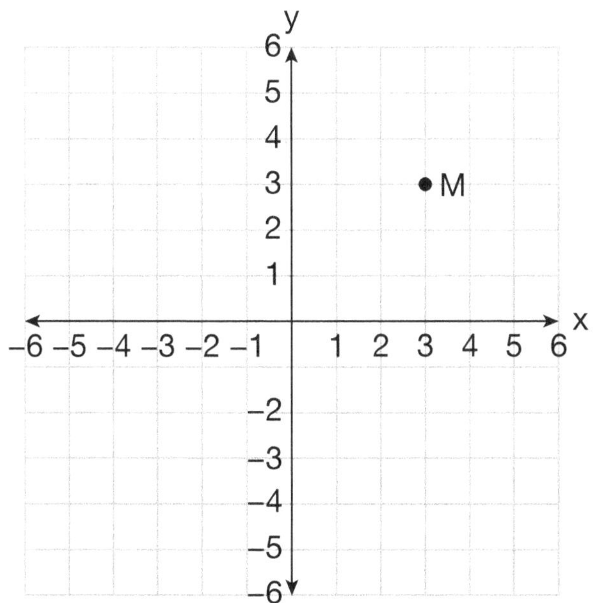

What would be the location of the image of M after it is reflected over the x–axis?

A) $M^{I} = (0, 3)$

B) $M^{I} = (3, 0)$

C) $M^{I} = (-3, 3)$

D) $M^{I} = (3, -3)$

2. What is the coordinate of the vertices of (–2, –4) when the vertices reflects across the y–axis.

A) (–2, –4)

B) (2, –4)

C) (2, 4)

D) (–2, 4)

3. Which expression describes the translation of a point from (–4, 5) to (6, –3)?

A) 10 units left and 8 units up

B) 10 units right and 8 units up

C) 10 units left and 8 units down

D) 10 units right and 8 units down

4.

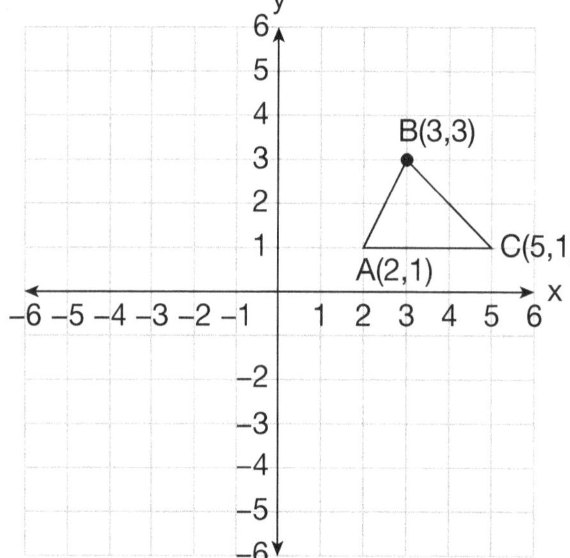

When △ABC is translated 5 units down, what is the apparent coordinate of A^{I}?

A) (2,1)

B) (–2,1)

C) (2,–4)

D) (4,2)

5. What is the coordinate of the vertices of (3,5) when the vertices reflects across the origin?

A) (3, 5)

B) (–3, 5)

C) (–3, –5)

D) (–5, –3)

CIRCLES

Radius: The distance from the center of the circle to the edge.

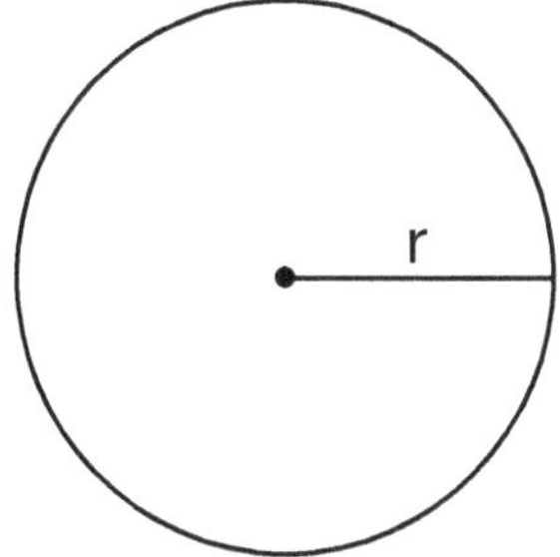

Chord: A line segment whose endpoints are on the circle.

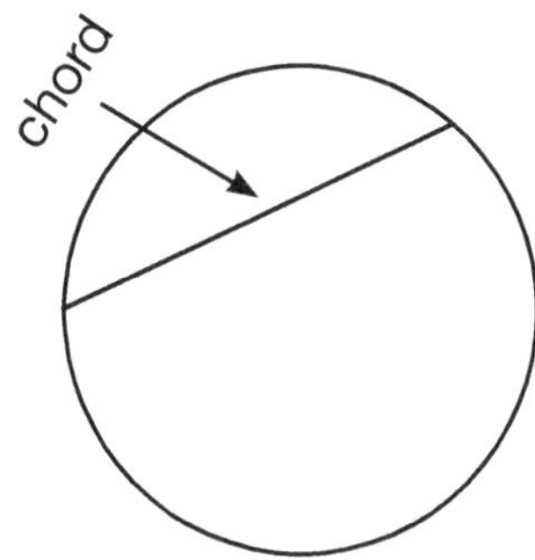

Diameter: A **chord** that passes through the center of the circle.

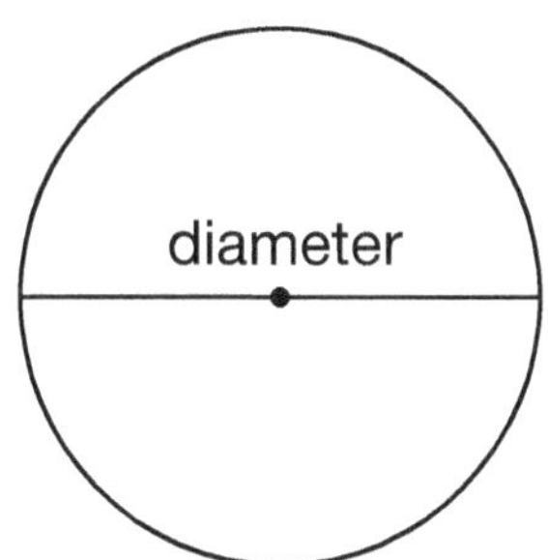

Tangent: A line which intersects a circle at exactly one point.

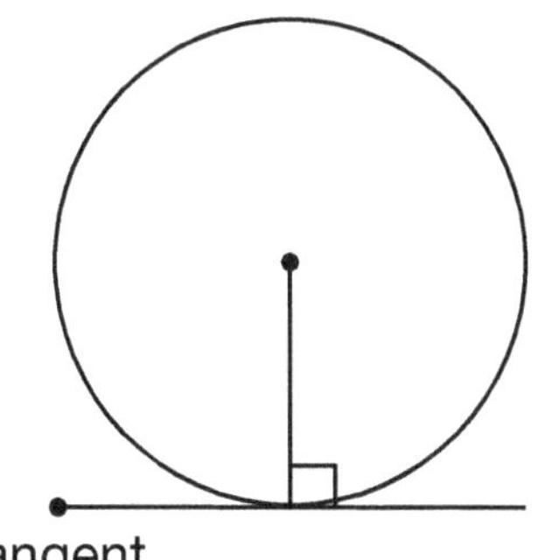

CIRCLES

Segment: A region of a circle which is "cut off" from the rest of the circle by a secant or a chord.

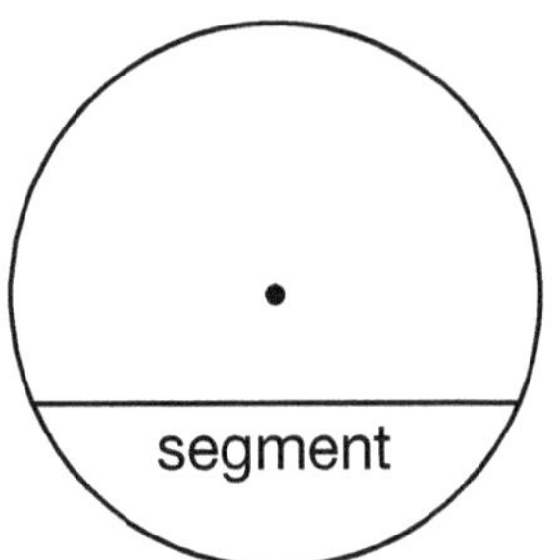

Area of a Circle:

$A = \pi r^2$

Circumference:

$C = 2\pi r$

Semi–Circle:

$A = \frac{1}{2}\pi r^2$

Area of the Sector:

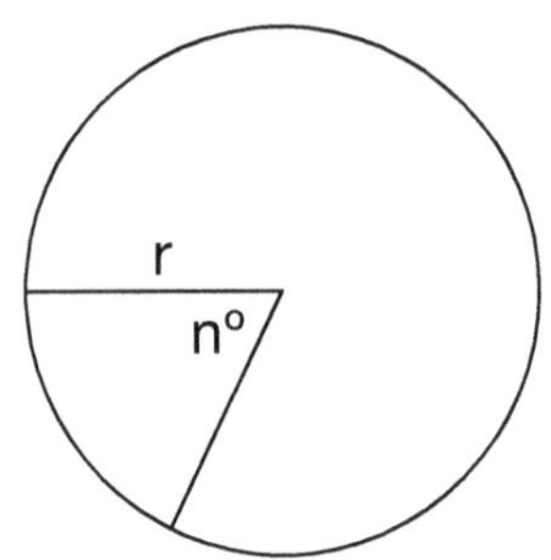

$A = \frac{n^\circ}{360^\circ}\pi r^2$

Equations of a Circle:

$(x - h)^2 + (y - k)^2 = r^2$

Central Angle:

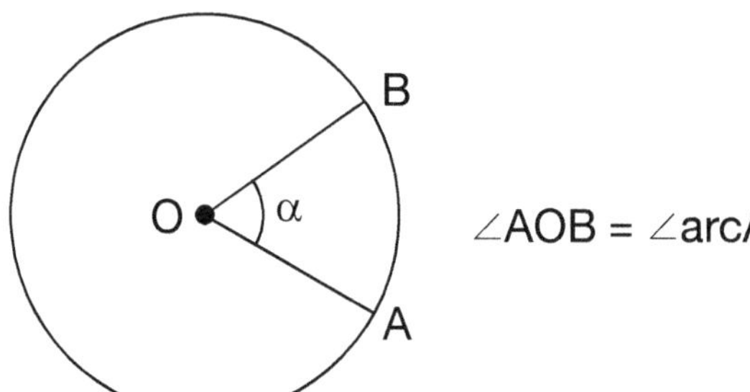

$\angle AOB = \angle arcAB$

Inscription of a Circle:

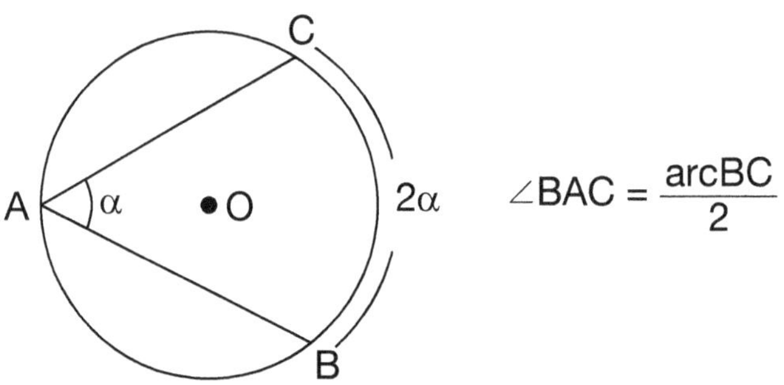

$\angle BAC = \frac{arcBC}{2}$

CIRCLES

Tangent Chord Angle:

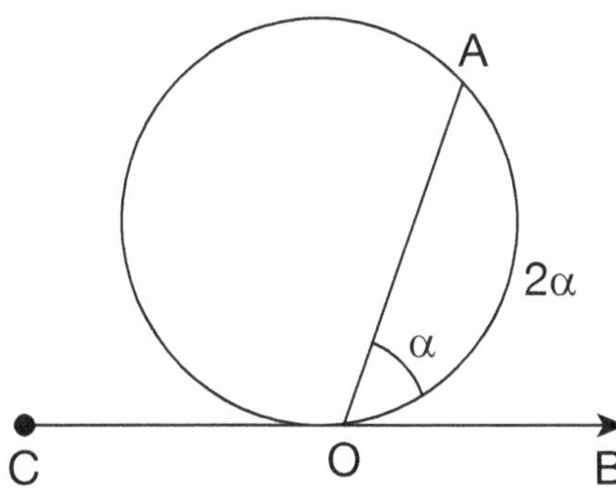

$$\angle AOC = \frac{\angle arc\ AO}{2}$$

Angle formed by two Intersecting Chords:

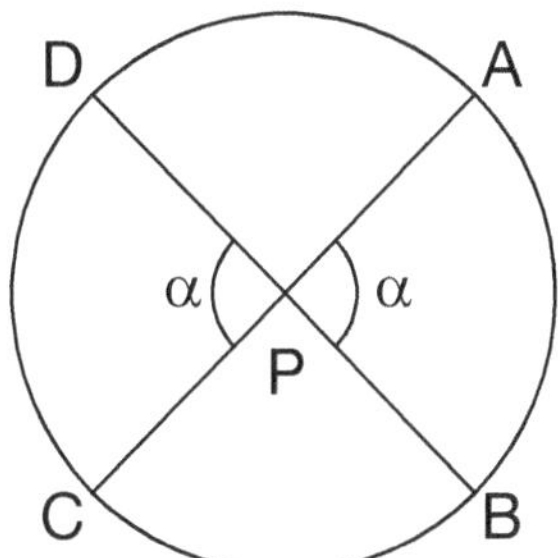

$$a = \angle DPC = (APB) = \frac{m(BA) + m(DC)}{2}$$

Angle formed outside of circle by Intersection:

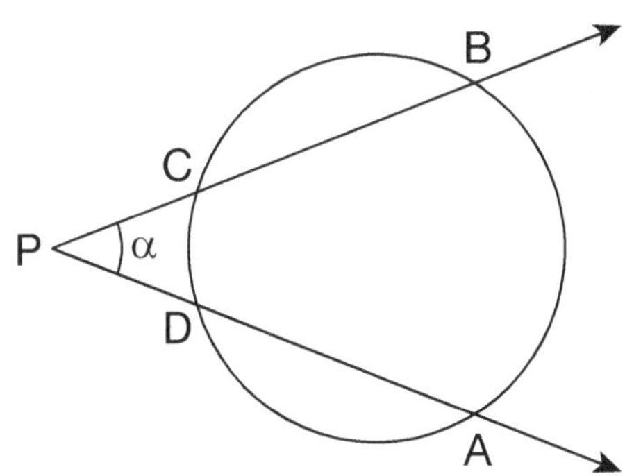

$$\angle BPA = \alpha = \frac{m(AB) - m(DC)}{2}$$

Length of an Arc formula:

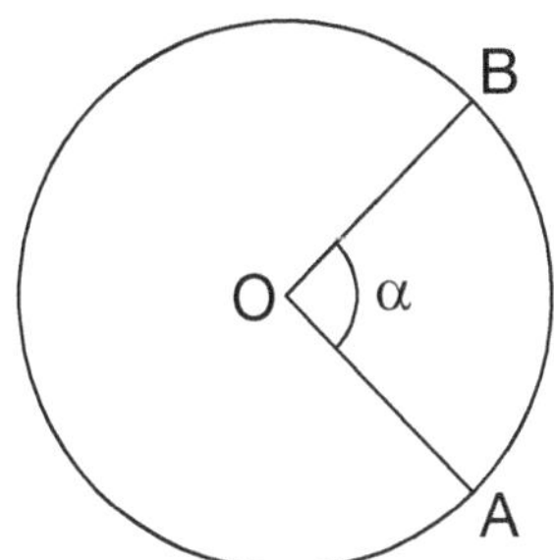

$$ArcAB = 2\pi r \frac{\alpha}{360°}$$

CIRCLES TEST

1. Find the circumference of the following circle.

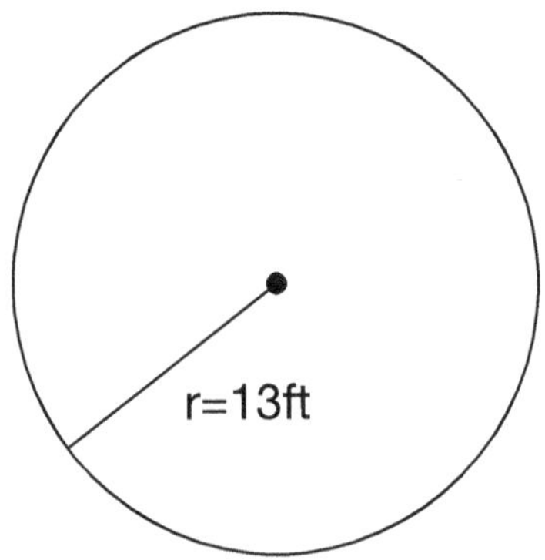

A) 13πft

B) 26πft

C) 36πft

D) 42πft

2. Find the area of the following circle.

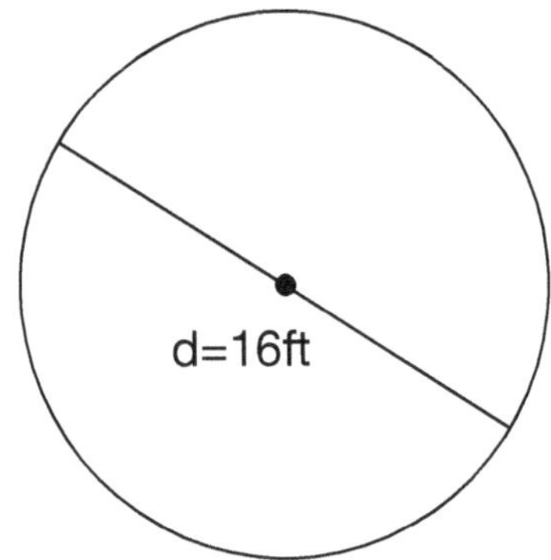

A) $6\pi ft^2$

B) $32\pi ft^2$

C) $48\pi ft^2$

D) $64\pi ft^2$

3. Using the two circles shown below, what is $\frac{\text{area of circle A}}{\text{area of circle B}}$?

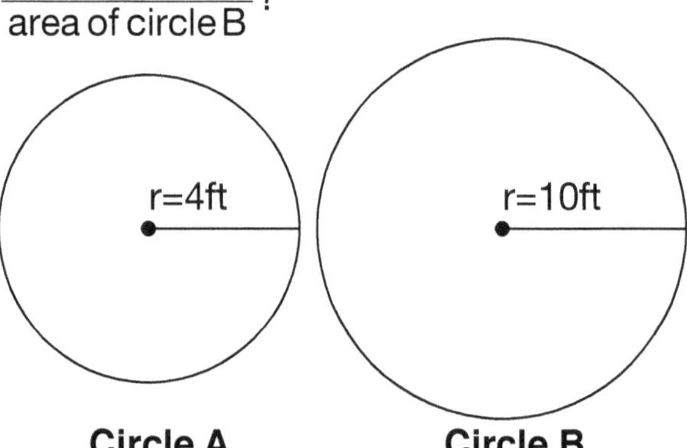

A) $\frac{4}{25}$

B) $\frac{1}{5}$

C) $\frac{3}{4}$

D) $\frac{4}{5}$

4. In the following figure, O is the center of the circle and the radius is 4cm. Find the area of the shaded part.

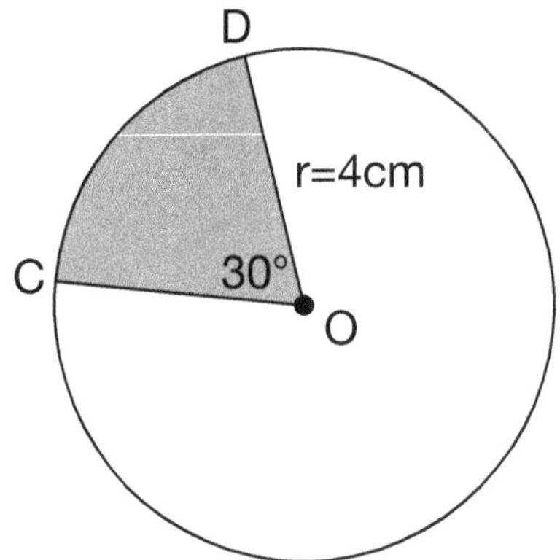

A) $\frac{3}{4}\pi cm^2$

B) $\frac{4}{3}\pi cm^2$

C) $4\pi cm^2$

D) $16\pi cm^2$

CIRCLES TEST

5. If the ratio of the circumference to the area of the circle is 4 to 6, what is the radius of the circle?

   A) 1

   B) 3

   C) 4

   D) 6

6. Using the two circles shown below, what is $\frac{\text{circumference of circle X}}{\text{circumference of circle Y}}$?

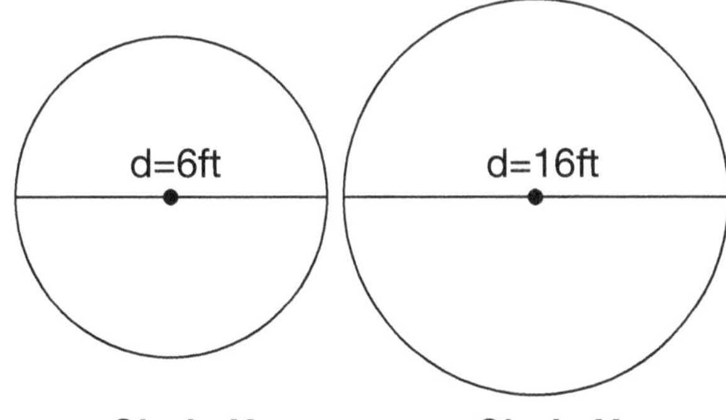

   A) $\frac{3}{8}$

   B) $\frac{3}{4}$

   C) $\frac{3}{2}$

   D) 3

7. In the following figure, what is the measure of a?

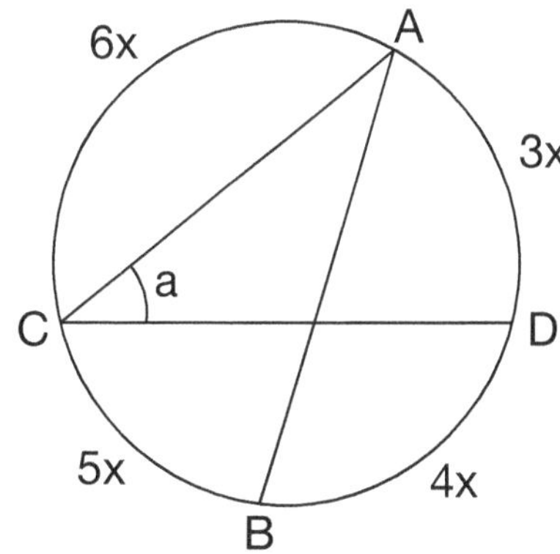

   A) 15°

   B) 25°

   C) 30°

   D) 45°

8. In the figure below, AB and CE are the diameters of the circles. What is the measure of x?

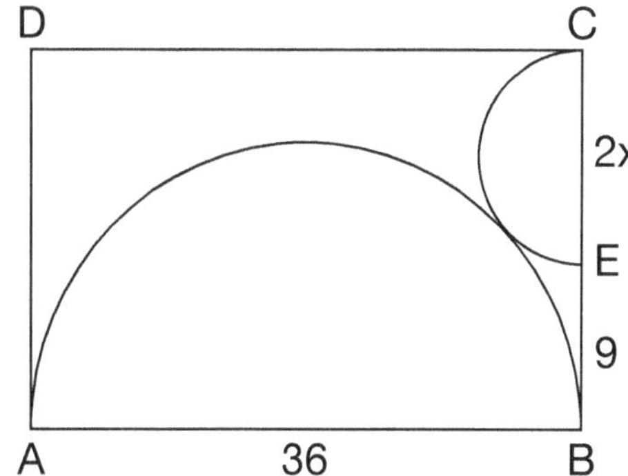

   A) 1

   B) 3

   C) 4.5

   D) 5

Formulas for right triangles

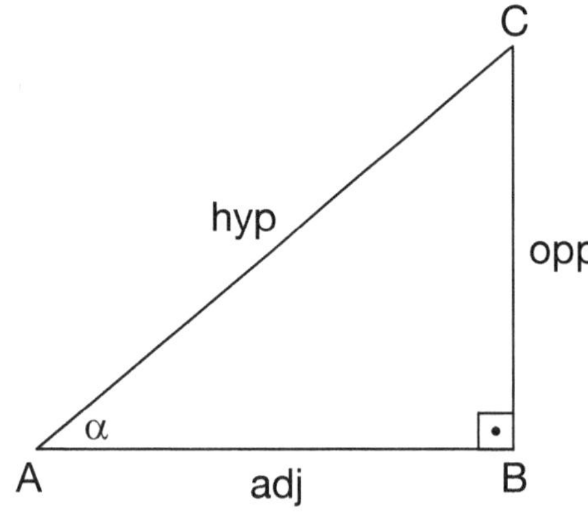

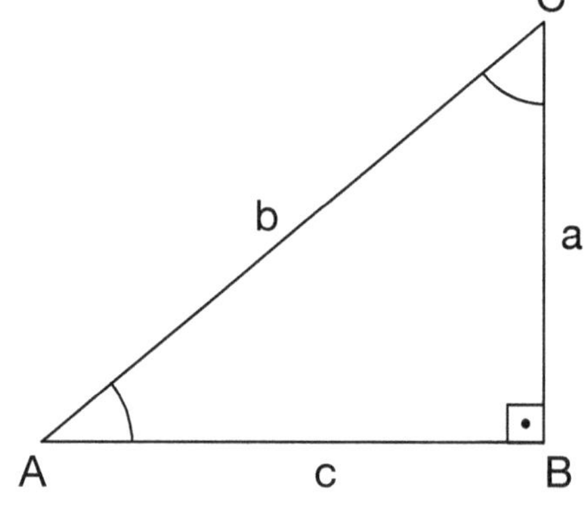

$$\sin\alpha = \frac{\text{opposite}}{\text{hypotenuse}}$$

$$\sin\widehat{A} = \frac{|BC|}{|AC|} = \frac{a}{b}$$

$$\sin\widehat{C} = \frac{|AB|}{|AC|} = \frac{c}{b}$$

$$\cos\alpha = \frac{\text{adjacent}}{\text{hypotenuse}}$$

$$\cos\widehat{A} = \frac{|AB|}{|AC|} = \frac{c}{b}$$

$$\cos\widehat{C} = \frac{|BC|}{|AC|} = \frac{a}{b}$$

$$\tan\alpha = \frac{\text{opposite}}{\text{adjacent}}$$

$$\tan\widehat{A} = \frac{|BC|}{|AB|} = \frac{a}{c}$$

$$\tan\widehat{C} = \frac{|AB|}{|BC|} = \frac{c}{a}$$

$$\cot a = \frac{\text{adjacent}}{\text{opposite}}$$

$$\cot\widehat{A} = \frac{|AB|}{|BC|} = \frac{c}{a}$$

$$\cot\widehat{C} = \frac{|BC|}{|AB|} = \frac{a}{c}$$

Convert radians to degrees

D: degrees

R: radians $\qquad \frac{D}{180} = \frac{R}{\pi}$

TRIGONOMETRY

Trigonometry on the unit circle

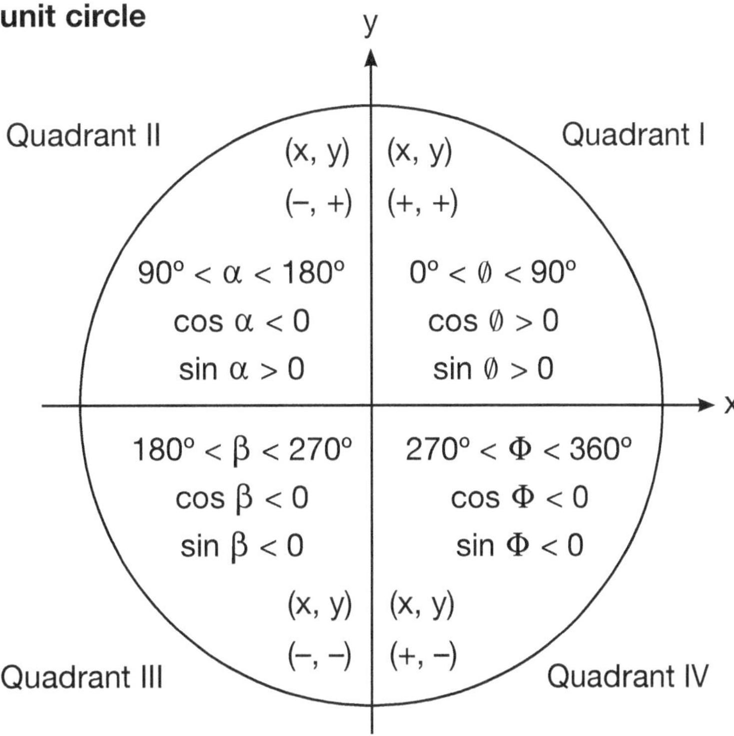

Trigonometric values of special angle

Degress	0°	30°	45°	60°	90°	180°	360°
Radians	0	$\frac{\pi}{6}$	$\frac{\pi}{4}$	$\frac{\pi}{3}$	$\frac{\pi}{2}$	π	2π
sin ∅	0	$\frac{1}{2}$	$\frac{1}{\sqrt{2}}$	$\frac{\sqrt{3}}{2}$	1	0	0
cos ∅	1	$\frac{\sqrt{3}}{2}$	$\frac{1}{\sqrt{2}}$	$\frac{1}{2}$	0	–1	1
tan ∅	0	$\frac{1}{\sqrt{3}}$	1	$\sqrt{3}$	Und	0	0
cot ∅	Und	$\sqrt{3}$	1	$\frac{1}{\sqrt{3}}$	0	Und	Und

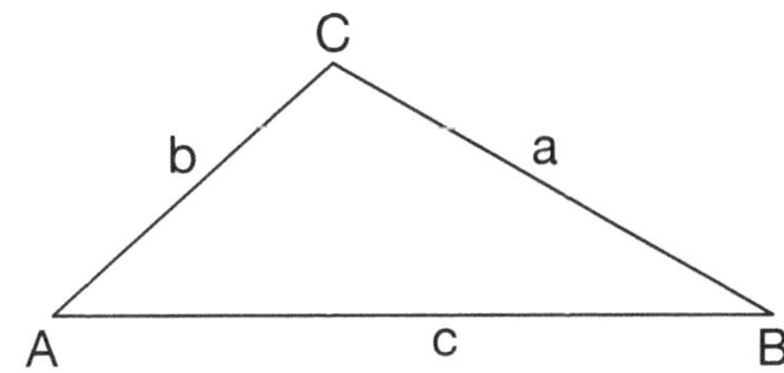

Law Of Cosines

$c^2 = a^2 + b^2 - 2 \cdot a \cdot b \cdot \text{Cosine } C$

Law Of Sines

$$\frac{\sin A}{a} = \frac{\sin B}{b} = \frac{\sin C}{c}$$

TRIGONOMETRY TEST

1. In the two right triangles below, if tan A = Cot D, find the value of EF.

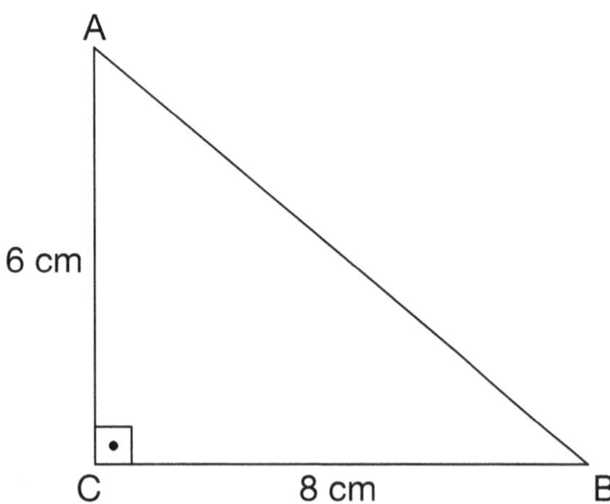

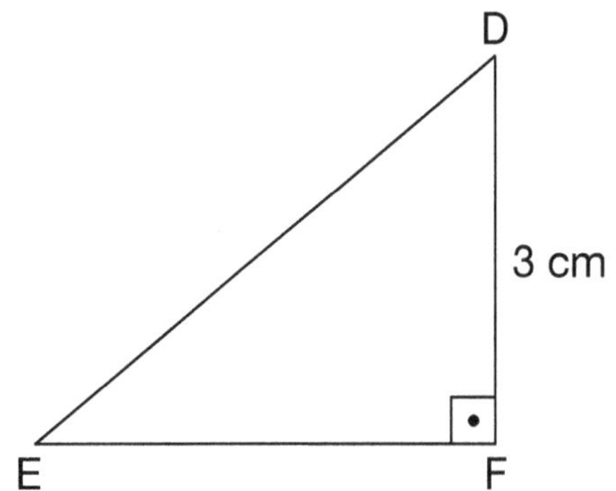

A) 4

B) 10

C) $\frac{9}{2}$

D) $\frac{9}{4}$

2. Convert 60° into radians measure. (Give your answer in terms of π)

A) $\frac{1}{3}\pi$

B) 3π

C) 2π

D) $\frac{1}{2}\pi$

American Math Academy

3. In the right triangle below, which of the following is correct?

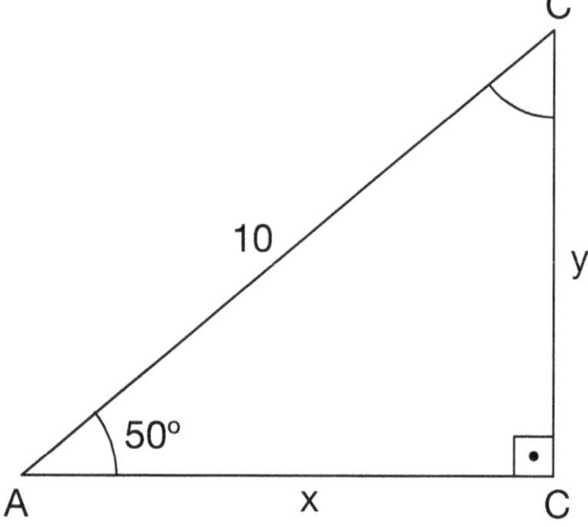

A) $\text{Sin}50° = \frac{x}{y}$

B) $\text{Tan}50° = \frac{y}{x}$

C) $\text{Cos}50° = \frac{10}{x}$

D) $\text{Cot}50° = \frac{10}{y}$

TRIGONOMETRY TEST

4. If $0° < x < 90°$ and $\sin x = \frac{3}{5}$, then find cotx.

A) $\frac{3}{4}$

B) $\frac{4}{3}$

C) $\frac{4}{5}$

D) 5

5. Find $\cos 30° \times \sin 60° \times \tan 60°$.

A) $\frac{\sqrt{3}}{3}$

B) $\frac{3}{4}$

C) $\sqrt{3}$

D) $\frac{3\sqrt{3}}{4}$

6. $\tan x + \frac{\cos x}{1+\sin x}$ what is the simplest form of the given equation?

A) secx

B) cosecx

C) cotx

D) sin2x

7. In a right triangle, one angle measures x°, where $\cos x° = \frac{5}{13}$.

What is $\tan(90° - x°)$?

A) $\frac{12}{13}$

B) $\frac{5}{12}$

C) $\frac{7}{12}$

D) $\frac{13}{12}$

American Math Academy

VOLUME

Right rectangular prism

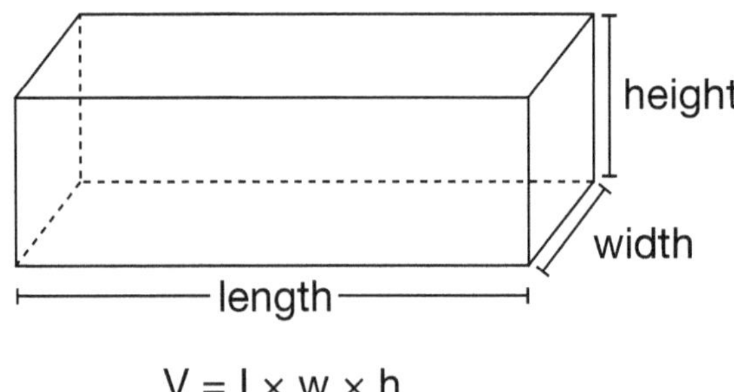

$V = l \times w \times h$

Cube

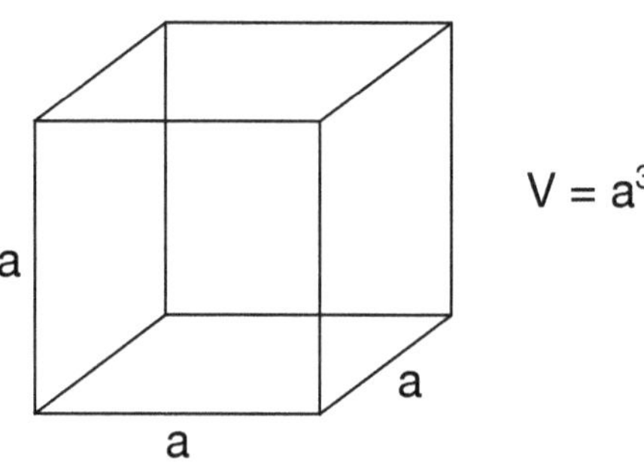

$V = a^3$

Cylinder

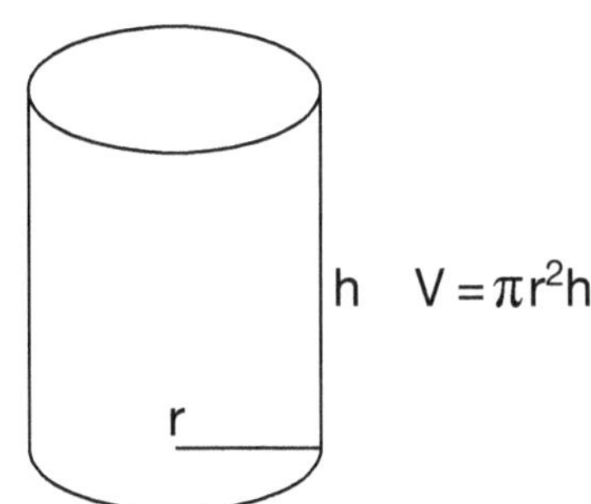

$V = \pi r^2 h$

Cone

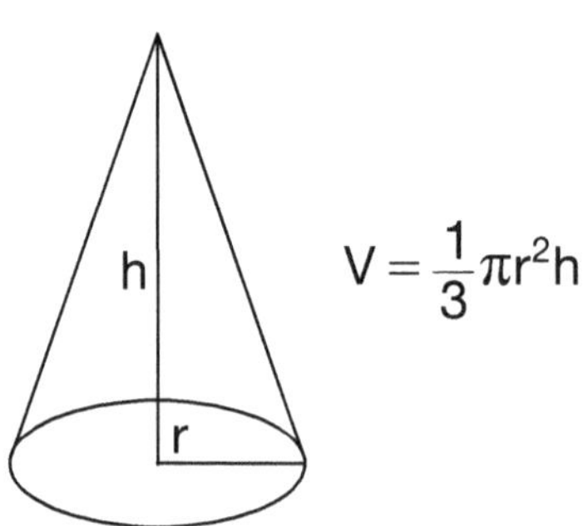

$V = \frac{1}{3}\pi r^2 h$

Pyramid

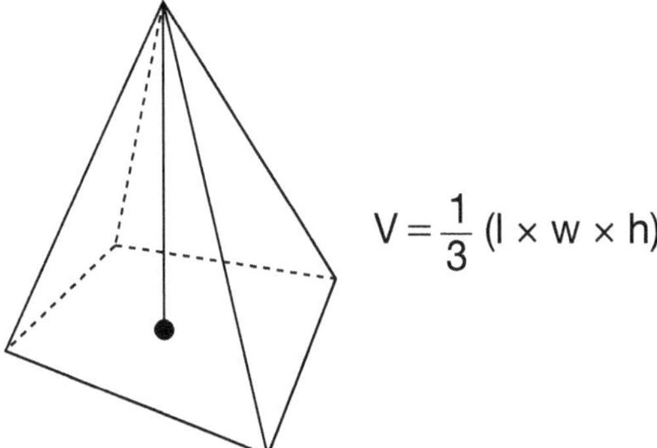

$V = \frac{1}{3}(l \times w \times h)$

Sphere

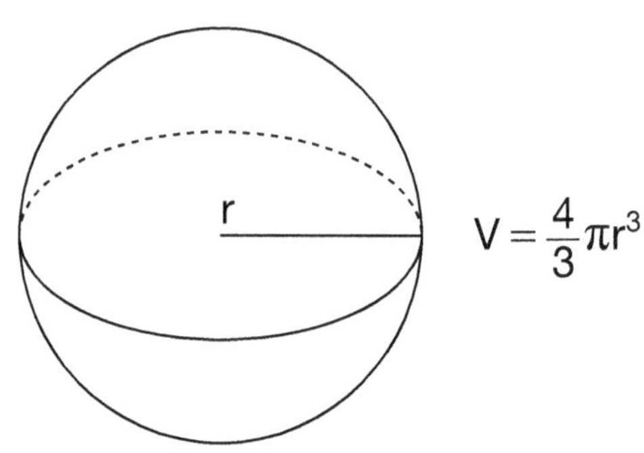

$V = \frac{4}{3}\pi r^3$

VOLUME TEST

1. A cone and a sphere have the same radius and volume. Find the radius r in terms of the height h.

   A) h

   B) 2h

   C) $\frac{h}{2}$

   D) $\frac{h}{4}$

2. The radius of a cylinder is increased by 25% and its height is decreased by 20%. What is the effect on the volume of cylinder?

   A) It is decreased by 50%

   B) It is decreased by 25%

   C) It is increased by 25%

   D) It is increased by 50%

3. If the volume of the cylinder is 72π cm^3, find the radius of the cylinder.

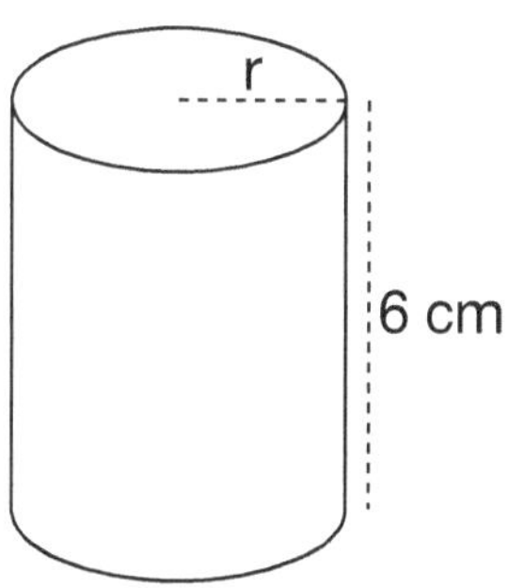

   A) $\sqrt{3}$ cm

   B) $2\sqrt{3}$ cm

   C) $4\sqrt{3}$ cm

   D) 6 cm

4. Find the volume of the following pyramid.

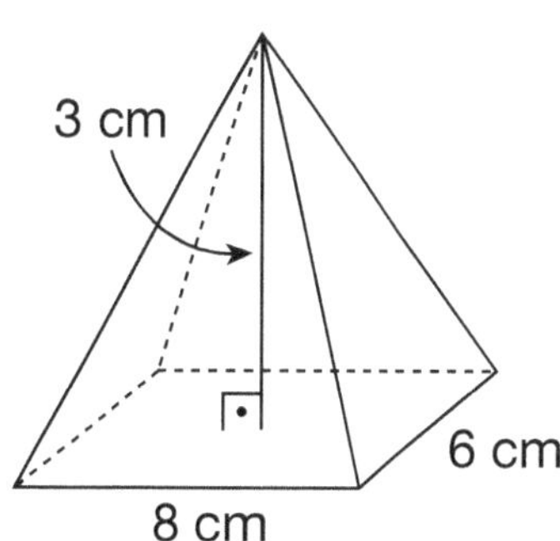

   A) 12 cm^3

   B) 24 cm^3

   C) 48 cm^3

   D) 60 cm^3

5. The volume of the following rectangular prism is 96 cm^3. What is value of h?

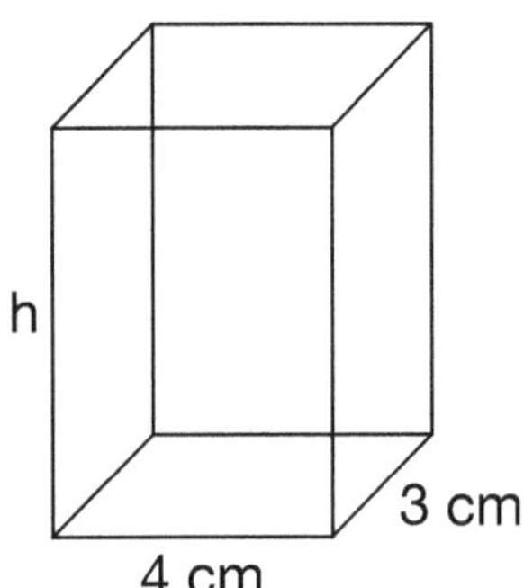

A) 4 cm

B) 8 cm

C) 12 cm

D) 16 cm

6. In the following figures, the volume of the cone and the cylinder are equal. What is the value of a?

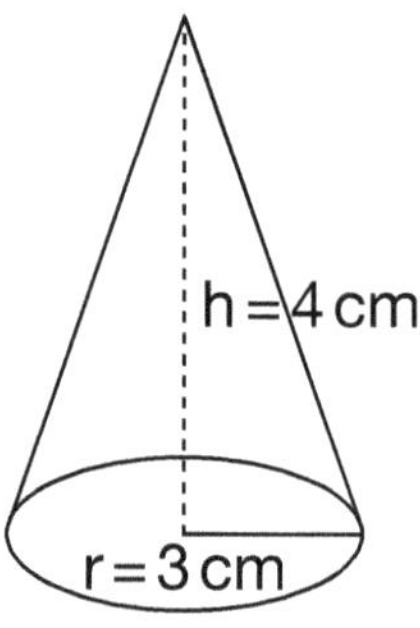

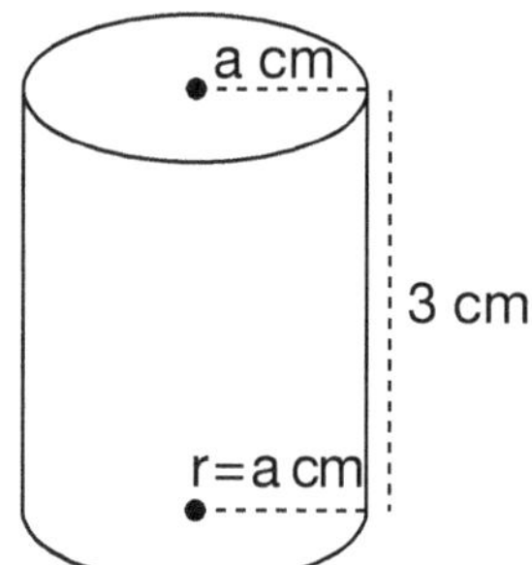

A) 1 cm

B) 2 cm

C) 3 cm

D) 4 cm

MIXED REVIEW TEST 1

1. If two complementary angles are in the ratio 4:5, what is the measure of the smaller angle?

   A) 36°

   B) 40°

   C) 45°

   D) 72°

2. If two parallel lines are cut by a transversal, what is the measure of each exterior angle if one interior angle measures 40°?

   A) 50°

   B) 140°

   C) 40°

   D) 90°

3. What is the scale factor of a dilation that transforms a square with a side length of 4 units to a square with a side length of 10 units?

   A) 2.5

   B) 1.5

   C) 2

   D) 6

4. If two triangles are congruent by SAS, what other congruence can be deduced?

   A) AAS

   B) ASA

   C) SSS

   D) None of the above

5. The diagonals of a rectangle are equal in length. If the length of the rectangle is 12 cm and the width is 9 cm, what is the length of one diagonal?

   A) 12 cm

   B) 15 cm

   C) 24 cm

   D) None of the above

6. What is the sum of the interior angles of a dodecagon (12-sided polygon)?

   A) 1800°

   B) 2160°

   C) 540°

   D) 360°

7. Two similar triangles have corresponding sides in the ratio 3:5. What is the ratio of their areas?

A) 3:5

B) 9:25

C) 9:5

D) 15:25

8. In a right triangle, if one acute angle measures 30°, what is the sine of the other acute angle?

A) 0.5

B) 1

C) $\frac{\sqrt{3}}{2}$

D) $\sqrt{\frac{2}{2}}$

9. What is the cosine of the angle θ if $\sin(\theta) = \frac{5}{13}$ and θ is an acute angle?

A) $\frac{12}{13}$

B) $\frac{5}{13}$

C) $-\frac{12}{13}$

D) $-\frac{5}{13}$

10. What is the area of a trapezoid with bases 10 cm and 15 cm, and height 8 cm?

A) 100 cm^2

B) 120 cm^2

C) 200 cm^2

D) 240 cm^2

11. A cylinder has a radius of 3 cm and a height of 10 cm. What is its total surface area?

(π = 3.14)

A) 244.92 cm^2

B) 198 cm^2

C) 282 cm^2

D) 318 cm^2

12. What is the volume of a cone with a radius of 4 cm and a height of 9 cm?

A) 48π cm^3

B) 144π cm^3

C) 36π cm^3

D) 72π cm^3

MIXED REVIEW TEST 1

13. The circumference of a circle is 31.4 cm. What is the radius of the circle?

A) 5 cm

B) 10 cm

C) 15 cm

D) 20 cm

14. The exterior angles of a polygon sum up to:

A) 90°

B) 180°

C) 270°

D) 360°

15. If the slope of line L_1 is 3 and line L_2 is perpendicular to L_1, what is the slope of line L_2?

A) -3

B) $\frac{1}{3}$

C) $-\frac{1}{3}$

D) 3

16. A point (x, y) is reflected over the x-axis. What are the coordinates of the reflected point?

A) (x, y)

B) (x, -y)

C) (-x, y)

D) (-x, -y)

17. If the two legs of one right triangle are congruent to the two legs of another right triangle, the triangles are congruent by:

A) SAS

B) ASA

C) SSS

D) HL

18. In a parallelogram, if one angle measures 60°, what is the measure of an adjacent angle?

A) 60°

B) 90°

C) 120°

D) 180°

19. In a regular hexagon, what is the measure of each interior angle?

A) 120°

B) 135°

C) 150°

D) 180°

20. Two similar triangles have a scale factor of 2:3. If the perimeter of the smaller triangle is 30 cm, what is the perimeter of the larger triangle?

A) 45 cm

B) 50 cm

C) 60 cm

D) 90 cm

21. In a right triangle, if the measures of the two legs are 7 cm and 24 cm, what is the measure of the hypotenuse?

A) 25 cm

B) 26 cm

C) 31 cm

D) 35 cm

22. In a right triangle, the tangent of an angle θ is $\frac{3}{4}$. What is the cosine of θ?

A) $\frac{4}{5}$

B) $\frac{3}{5}$

C) $\frac{5}{4}$

D) $\frac{5}{3}$

23. What is the area of a sector of a circle with a radius of 10 cm and a central angle of 90°?

A) 25π cm²

B) 50π cm²

C) 100π cm²

D) 10π cm²

24. A rectangular prism has dimensions 4 cm by 5 cm by 6 cm. What is its total surface area?

A) 94 cm²

B) 104 cm²

C) 144 cm²

D) 148 cm²

MIXED REVIEW TEST 2

1. A line segment is determined by how many points?

A) One

B) Two

C) Three

D) Infinite

2. In a Cartesian coordinate system, the point (0,0) is referred to as the:

A) Origin

B) X-intercept

C) Y-intercept

D) Endpoint

3. If two lines are parallel and a third line intersects them, the third line is called a:

A) Perpendicular line

B) Transversal

C) Segment

D) Ray

4. What is the slope of a line perpendicular to a line with a slope of 4?

A) 4

B) -4

C) $\frac{1}{4}$

D) $-\frac{1}{4}$

5. Which transformation does not change the size or shape of the object?

A) Reflection

B) Rotation

C) Translation

D) Dilation

6. What is the name of the transformation that reduces or enlarges a figure?

A) Translation

B) Reflection

C) Dilation

D) Rotation

7. Which postulate states that two triangles are congruent if two sides and the included angle of one triangle are congruent to two sides and the included angle of another triangle?

A) SAS

B) ASA

C) SSS

D) AAS

8. In congruent triangles, corresponding angles are:

A) Complementary

B) Supplementary

C) Equal

D) None of the above

9. What is the sum of interior angles of a quadrilateral?

A) 180°

B) 270°

C) 360°

D) 450°

10. A quadrilateral with one pair of parallel sides is called a:

A) Parallelogram

B) Rectangle

C) Trapezoid

D) Square

11. What is the sum of interior angles of a hexagon?

A) 540°

B) 720°

C) 900°

D) 1080°

12) In a regular polygon, all sides and all angles are:

A) Equal

B) Unequal

C) Complementary

D) Supplementary

MIXED REVIEW TEST 2

13. Two triangles are similar if they have:

A) Equal sides

B) Equal angles

C) Corresponding sides are proportional

D) None of above

14. The ratio of the perimeters of two similar triangles is 2:3. If the smaller triangle has a perimeter of 30 cm, what is the perimeter of the larger triangle?

A) 45 cm

B) 60 cm

C) 90 cm

D) 120 cm

15. In a right triangle, the side opposite the right angle is called the:

A) Hypotenuse

B) Adjacent

C) Opposite

D) Median

16. In a right-angled triangle, if one angle measures 30°, what is the measure of the other acute angle?

A) 30°

B) 45°

C) 60°

D) 90°

17. In a right triangle, sin(θ) is equal to:

A) Opposite/Adjacent

B) Adjacent/Hypotenuse

C) Opposite/Hypotenuse

D) Hypotenuse/Opposite

18. What is the cosine of a 60° angle?

A) $\frac{1}{2}$

B) $\frac{\sqrt{3}}{2}$

C) 1

D) $\frac{\sqrt{2}}{2}$

MIXED REVIEW TEST 2

19. What is the formula for the area of a trapezoid?

A) (Sum of bases) × Height

B) $\frac{1}{2}$ × (Sum of bases) × Height

C) Base × Height

D) $\frac{1}{2}$ × Base × Height

20. The area of a rectangle is 48 cm^2, and the length is 8 cm. What is the width?

A) 4 cm

B) 6 cm

C) 8 cm

D) 12 cm

21. What is the formula for the surface area of a cylinder?

A) $2\pi r(h + r)$

B) $2\pi rh + \pi r^2$

C) $2\pi r^2 + 2\pi rh$

D) $\pi r^2 h$

22. A rectangular prism has dimensions 3 cm, 4 cm, and 5 cm. What is the total surface area?

A) 94 cm^2

B) 104 cm^2

C) 114 cm^2

D) 124 cm^2

23. What is the formula for the volume of a sphere?

A) $\frac{4}{3}\pi r^3$

B) $\frac{2}{3}\pi r^3$

C) $\pi r^2 h$

D) πr^2

24. A cylinder has a radius of 3 cm and a height of 4 cm. What is the volume?

A) 36π cm^3

B) 48π cm^3

C) 60π cm^3

D) 72π cm^3

MIXED REVIEW TEST 3

1. If triangle ABC is similar to triangle DEF, and the ratio of the lengths of AB to DE is 3:5, what is the ratio of the areas of the two triangles?

A) 3:5

B) 5:3

C) 9:25

D) 25:9

2. If two lines are cut by a transversal so that alternate interior angles are congruent, the lines are:

A) Perpendicular

B) Parallel

C) Coincident

D) Skew

3. A rotation of 90° counterclockwise about the origin takes the point (x, y) to:

A) (y, -x)

B) (-y, x)

C) (-x, -y)

D) (x, y)

4. If two triangles are congruent, then their corresponding parts are:

A) Congruent

B) Supplementary

C) Complementary

D) Proportional

5. The diagonals of a rectangle are:

A) Perpendicular

B) Congruent

C) Both A and B

D) Neither A nor B

6. The sum of the exterior angles of a polygon, one at each vertex, is:

A) 180°

B) 270°

C) 360°

D) 540°

MIXED REVIEW TEST 3

7. Two triangles are similar if they have:

A) Two pairs of congruent angles

B) Two pairs of congruent sides

C) Three pairs of proportional sides

D) Three pairs of congruent angles

8. In a 45-45-90 right triangle, the ratio of the length of the hypotenuse to the length of a leg is:

A) 1:1

B) $1:\sqrt{2}$

C) $\sqrt{2}:1$

D) 2:1

9. $\sin(\theta) = \frac{3}{5}$, and θ is acute, then cos(θ) equals:

A) $\frac{4}{5}$

B) $\frac{5}{3}$

C) $\frac{5}{4}$

D) $\frac{3}{4}$

10. The area of a rhombus with diagonals of lengths 10 cm and 24 cm is:

A) 120 cm^2

B) 240 cm^2

C) 480 cm^2

D) 960 cm^2

11. The total surface area of a sphere with radius 5 cm is:

A) 25π cm^2

B) 50π cm^2

C) 100π cm^2

D) 200π cm^2

12. The volume of a cone with radius 3 cm and height 4 cm is:

A) 12π cm^3

B) 24π cm^3

C) 36π cm^3

D) 48π cm^3

MIXED REVIEW TEST 3

13. The equation of a circle with center (2, -3) and radius 5 is:

A) $(x - 2)^2 + (y + 3)^2 = 25$

B) $(x + 2)^2 + (y - 3)^2 = 25$

C) $(x - 2)^2 + (y + 3)^2 = 5$

D) $(x + 2)^2 + (y - 3)^2 = 5$

14. The measure of an angle formed by two tangent lines to a circle that intersect outside the circle is:

A) Half the difference of the intercepted arcs

B) Half the sum of the intercepted arcs

C) Equal to the difference of the intercepted arcs

D) Equal to the sum of the intercepted arcs

15. The slope of a line perpendicular to the line $3x - 4y = 73$ is:

A) $-\frac{3}{4}$

B) $\frac{4}{3}$

C) $-\frac{4}{3}$

D) $\frac{3}{4}$

16. The matrix $\begin{pmatrix} 0 & -1 \\ 1 & 0 \end{pmatrix}$ represents a:

A) Reflection across the x-axis

B) Reflection across the y-axis

C) 90° rotation counterclockwise

D) 90° rotation clockwise

17. If two triangles are congruent, their corresponding medians are:

A) Parallel

B) Perpendicular

C) Congruent

D) None of the above

18. In a parallelogram, opposite angles are:

A) Complementary

B) Supplementary

C) Equal

D) None of the above

MIXED REVIEW TEST 3

19. The interior angle of a regular pentagon is:

A) 108°

B) 120°

C) 135°

D) 150°

20. Two triangles are similar if their corresponding sides are in proportion and their corresponding angles are:

A) Equal

B) Supplementary

C) Complementary

D) None of the above

21. The Pythagorean Theorem is applicable to:

A) All triangles

B) Only right-angled triangles

C) Only equilateral triangles

D) Only isosceles triangles

22. The tangent of angle θ in a right-angled triangle is the ratio of:

A) Opposite side to Adjacent side

B) Adjacent side to Hypotenuse

C) Opposite side to Hypotenuse

D) Hypotenuse to Opposite side

23. The area of an equilateral triangle with side length a is:

A) $\frac{a^2\sqrt{3}}{4}$

B) $\frac{a^2}{2}$

C) $a^2\sqrt{3}$

D) $\frac{a^2}{4}$

24. A cylinder has a radius of r and a height of h. The total surface area of the cylinder is:

A) $2\pi r(h + r)$

B) $\pi r^2 + 2\pi r h$

C) $2\pi r^2 + \pi r h$

D) $\pi r^2 h$

MIXED REVIEW TEST 4

1. If triangle ABC is similar to triangle DEF, and the length of side AB is 9 cm,

What is the length of side DE if the ratio of the corresponding side lengths is 5:6?

A) 10.8 cm

B) 12 cm

C) 15 cm

D) 16.2 cm

2. An arc that measures 90° is what fraction of a circle?

A) $\frac{1}{4}$

B) $\frac{1}{2}$

C) $\frac{1}{3}$

D) $\frac{1}{6}$

3. What is the length of an arc that subtends an angle of 60° at the center of a circle with radius 10 cm?

A) 10π cm

B) 5π cm

C) π cm

D) $\frac{10\pi}{3}$ cm

4. How many radians are there in a full circle?

A) π

B) 2π

C) 3π

D) 4π

5. What is the length of an arc subtended by an angle of $\frac{\pi}{3}$ radians in a circle with radius 6 cm?

A) 2 cm

B) 2π cm

C) 4 cm

D) 4π cm

6. The area of a sector with an angle of 45° in a circle with radius 8 cm is:

A) 8π cm²

B) 16π cm²

C) 32π cm²

D) 64π cm²

7. An inscribed angle subtends a diameter of a circle. What is the measure of the inscribed angle?

A) 45°

B) 60°

C) 90°

D) 180°

8. In a circle, an inscribed equilateral triangle has a side length of 6 cm. What is the radius of the circle?

A) 3 cm

B) $3\sqrt{3}$ cm

C) $2\sqrt{3}$ cm

D) $4\sqrt{3}$ cm

9. Two tangents drawn from an external point to a circle are:

A) Parallel

B) Congruent

C) Perpendicular to each other

D) None of the above

10. An equilateral triangle is inscribed in a circle with radius 12 cm. What is the area of the triangle?

A) $18\sqrt{3}$ cm²

B) $48\sqrt{3}$ cm²

C) $96\sqrt{3}$ cm²

D) $108\sqrt{3}$ cm²

11. What is the standard equation of a circle with a center at (2, 3) and a radius of 5?

A) $(x - 2)^2 + (y - 3)^2 = 25$

B) $(x + 2)^2 + (y + 3)^2 = 25$

C) $(x - 2)^2 + (y - 3)^2 = 5$

D) $(x + 2)^2 + (y + 3)^2 = 5$

12. The equaiton $x^2 + y^2 + 4x - 6y - 12 = 0$ is the expanded form of the equation of a circle.
What is the center of the circle?

A) (–2, 3)

B) (2, –3)

C) (–2, –3)

D) (2, 3)

MIXED REVIEW TEST 4

13. Identify the slope and y-intercept of the line represented by the equation $y = 2x + 3$.

A) Slope = 2, y-intercept = 3

B) Slope = 3, y-intercept = 2

C) Slope = 2, y-intercept = –3

D) Slope = –3, y-intercept = 2

14. What is the truth value of the statement

$p \wedge -p$?

A) True

B) False

C) Indeterminate

D) None of the above

15. The contrapositive of the statement $p \longrightarrow q$ is:

A) $q \longrightarrow p$

B) $-p \longrightarrow -q$

C) $-q \longrightarrow -p$

D) $p \wedge q$

16. The biconditional statement $p \longleftrightarrow q$ is true when:

A) p and q are both true or both false

B) p is true and q is false

C) p is false and q is true

D) p and q are always true

17. Which of the following transformations preserves both distance and angle measures?

A) Translation

B) Rotation

C) Reflection

D) All of the above

18. What is the result of translating point (3, 4) 5 units to the right and 3 units up?

A) (8, 7)

B) (8, 1)

C) (-2, 1)

D) (-2, 7)

MIXED REVIEW TEST 4

19. What is the result of rotating point (2, 3) 90° counterclockwise about the origin?

A) (-3, 2)

B) (3, -2)

C) (-2, -3)

D) (3, 2)

20. What is the result of reflecting point (4, 5) over the x-axis?

A) (4, -5)

B) (-4, 5)

C) (-4, -5)

D) (5, 4)

21. Which of the following is NOT a rigid transformation?

A) Translation

B) Rotation

C) Reflection

D) Dilation

22. A dilation with a scale factor of 2 centered at the origin transforms point (1, 2) to:

A) (2, 4)

B) (3, 4)

C) (0.5, 1)

D) (-1, -2)

23. An object has rotational symmetry if it can be rotated less than 360° and still look the same as:

A) Its reflection

B) Its original position

C) Its translation

D) Its dilation

FINAL TEST

1. If $|AC| = |CD|$, $\angle(ACD) = \angle(DCB)$ and $\angle(BAC) = 70°$

Find x.

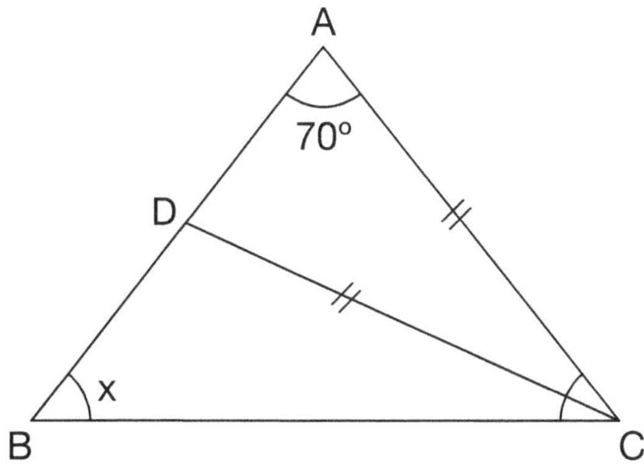

A) 30°

B) 40°

C) 50°

D) 60°

2. Find the value of x in the diagram.

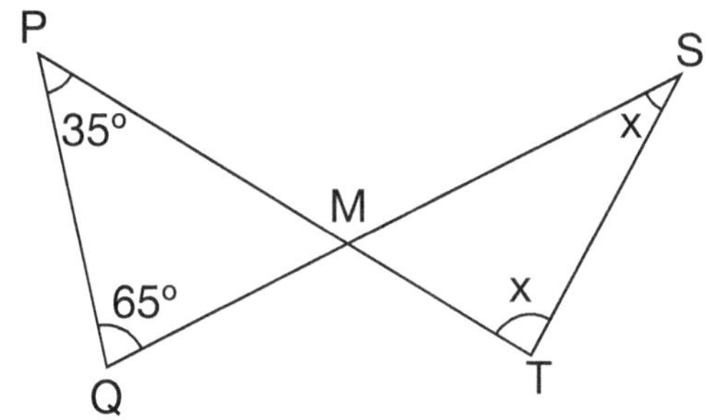

A) 25°

B) 35°

C) 45°

D) 50°

3.

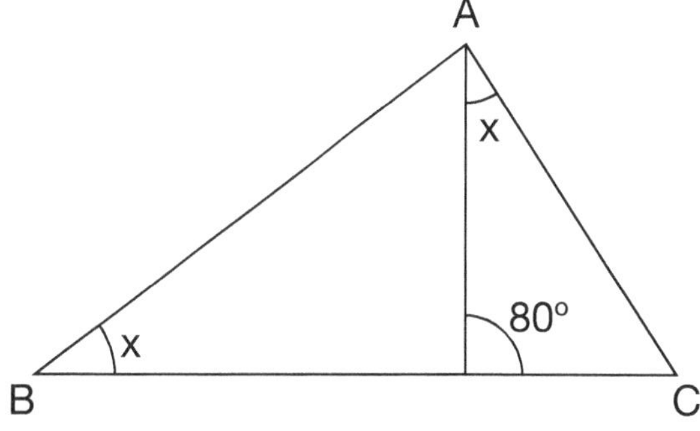

Find the angle of $\angle(BAC)$.

A) 60°

B) 70°

C) 80°

D) 90°

4.

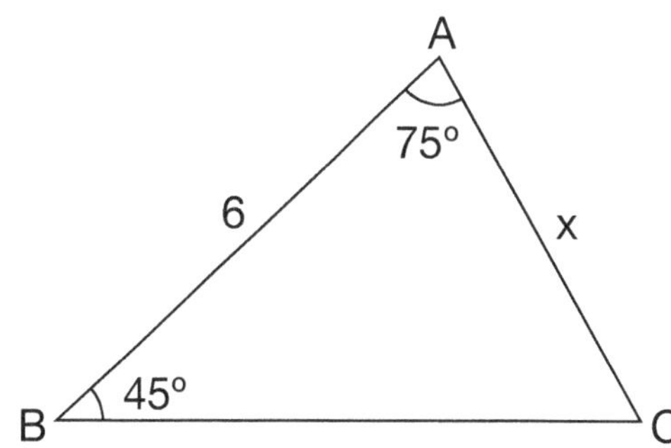

What is the value of x?

A) $\sqrt{6}$

B) $3\sqrt{2}$

C) $2\sqrt{6}$

D) 6

5.

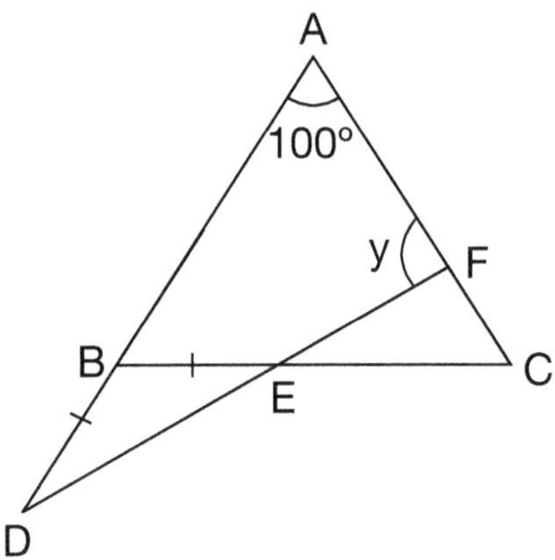

From the above figure if $|\overline{AB}| = |\overline{AC}|$ and $|\overline{BD}| = |\overline{BE}|$, what is the value of y?

A) 15°

B) 30°

C) 45°

D) 60°

6.

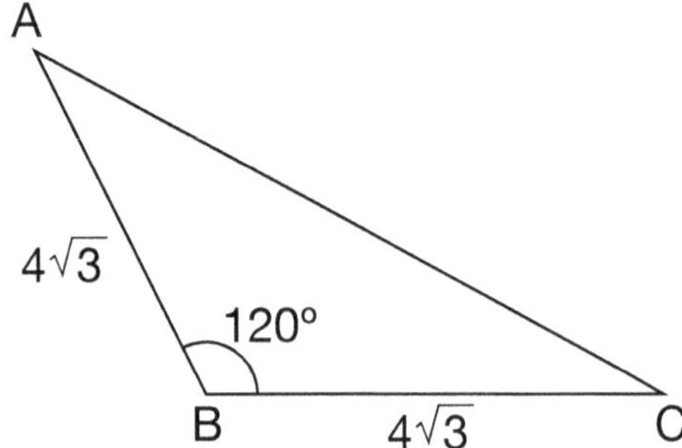

From the figure, what is the area of △(ABC)?

A) $12\sqrt{3}$

B) $8\sqrt{3}$

C) $9\sqrt{3}$

D) $18\sqrt{3}$

7.

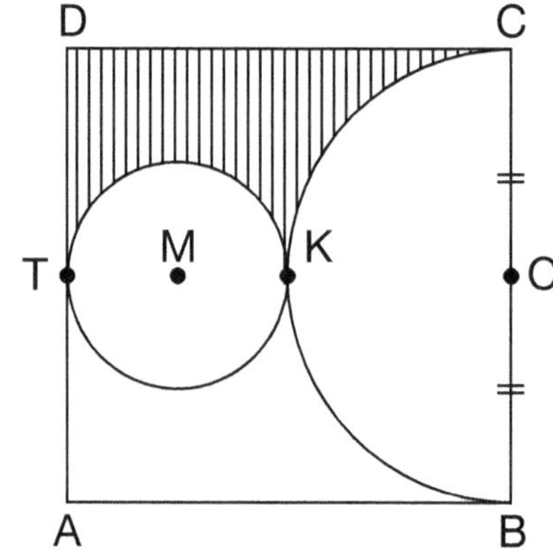

In the following figures, M and O are the center of the circles, ABCD is a square and AB = 4 feet. What is the area of the shaded part?

A) $8 - \frac{3\pi}{2}$

B) $4 - \frac{3\pi}{2}$

C) $\frac{3\pi}{2} - 2$

D) $\frac{3\pi}{2}$

8. In the following figure, what is the measure of x?

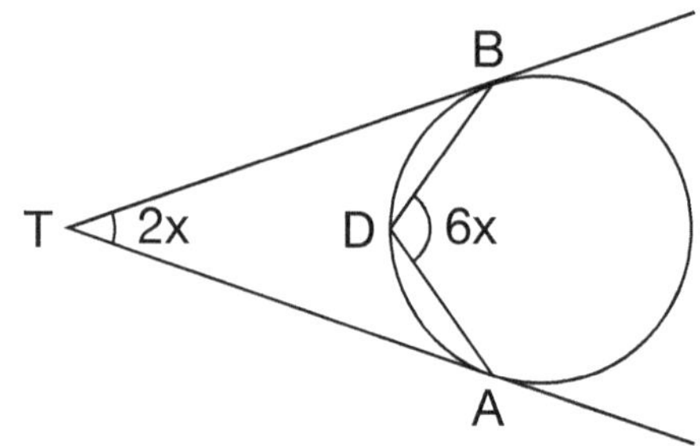

A) 6

B) 12

C) 18

D) 24

9. The angles of a triangle are in the ratio of 4 : 5 : 9. What is the degree measure of the smallest angle?

A) 40°

B) 30°

C) 25°

D) 20°

10. In the following circle AB = 3, BC = 6 and, B is the center of both circles.
Find the shaded area.

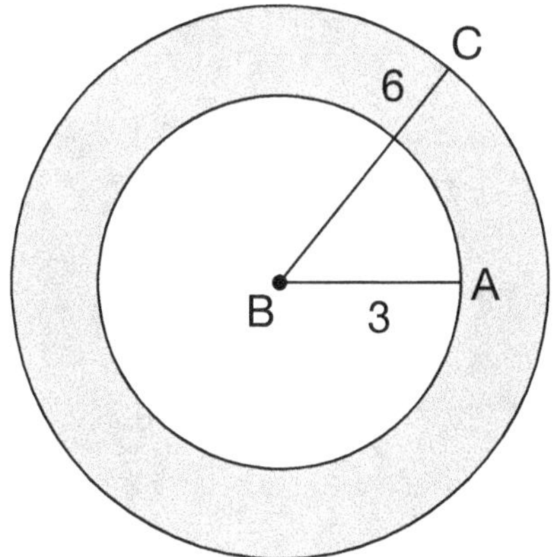

A) 9π

B) 12π

C) 18π

D) 27π

11. Find the equation of the circle centered at (3,–2) with a radius of 5.

A) $(x-3)^2+(y-2)^2=5$

B) $(x-3)^2+(y+2)^2=5$

C) $(x+3)^2+(y+2)^2=5$

D) $(x-3)^2+(y+2)^2=25$

12. In the following figure O is the center of the circle. Find the area of the shaded part.

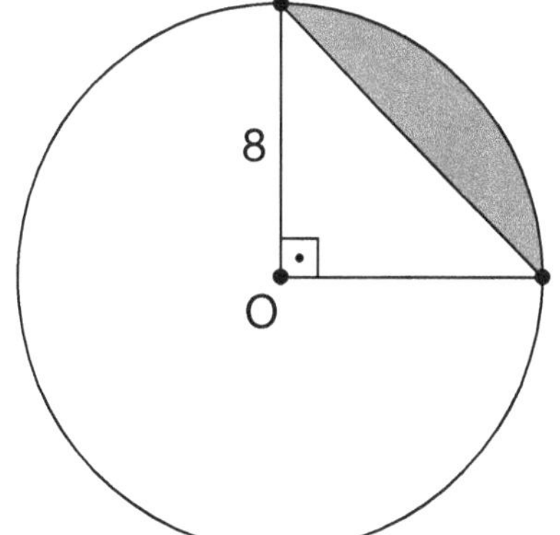

A) $16-32\pi$

B) $16\pi-32$

C) 32π

D) 16π

13. If $0° < x < 90°$ and $\sin x = \frac{3}{5}$, then find $\cot x$?

A) $\frac{3}{4}$

B) $\frac{4}{3}$

C) $\frac{4}{5}$

D) 5

14. Which of the following equations passes through the coordinates (1, 2) and (–1, –2)?

A) $y = x$

B) $y = 2x$

C) $y = \frac{1}{2}x + 1$

D) $y = 2x - 1$

SAMPLE TEST SOLUTIONS

1. **Solution:**

$2x - 5^\circ + 8x - 15^\circ + 90^\circ = 180^\circ$

$10x - 20^\circ = 90^\circ$

$+20^\circ \quad 20^\circ$

$10x = 110^\circ$

$x = \frac{110}{10} = 11^\circ$

Correct Answer : D

2. **Solution:**

$5x + 2x + 60^\circ + 90^\circ = 360^\circ$

$7x + 150^\circ = 360^\circ$

$-150^\circ \quad -150^\circ$

$7x = 210^\circ$

$x = 30^\circ$

Correct Answer : B

3. **Solution:**

$3x + 5 + 2x + 10 = 30$

$5x + 15 = 30$

$-15 \quad -15$

$5x = 15$

$x = 3$

Correct Answer : C

4. **Solution:**

$m\angle DEF = m\angle FEG,$

$3x - 10 = 2x + 15$

$-2x \quad -2x$

$x - 10 = 15, x = 25$

Correct Answer : D

5. **Solution:**

$(x + 5)^\circ + (2x + 10)^\circ + 90^\circ = 180^\circ$

$3x + 15^\circ + 90^\circ = 180$

$3x + 105^\circ = 180^\circ$

$-105^\circ \quad -105^\circ$

$3x = 75^\circ$

$x = 25^\circ$

Correct Answer : B

6. **Solution:**

$3x + 5^\circ + 8x + 10^\circ = 5x + 75^\circ$

$11x + 15^\circ = 5x + 75^\circ$

$-15^\circ \quad -15^\circ$

$11x = 5x + 60^\circ$

$-5x = -5x$

$6x = 60^\circ$

$x = 10^\circ$

Correct Answer : D

American Math Academy

7. Solution:

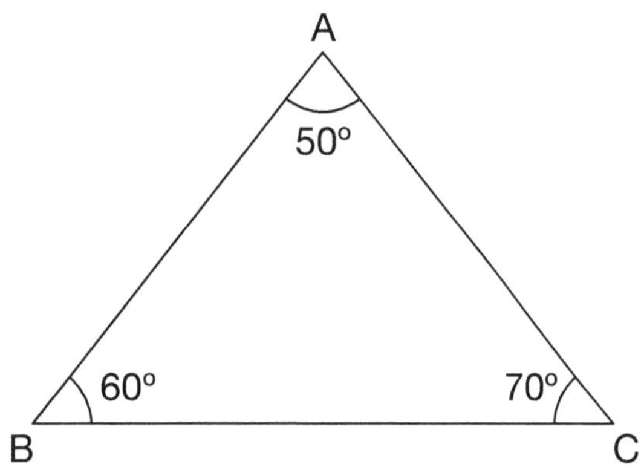

The shortest side is always opposite the smallest interior angle.

Sides of shortest to longest:

$\overline{AC} = 60°$, $\overline{BC} = 50°$, and $\overline{AB} = 70°$

$\overline{BC} \angle \overline{AC} \angle \overline{AB}$

Correct Answer : D

8. Solution:

Rule:

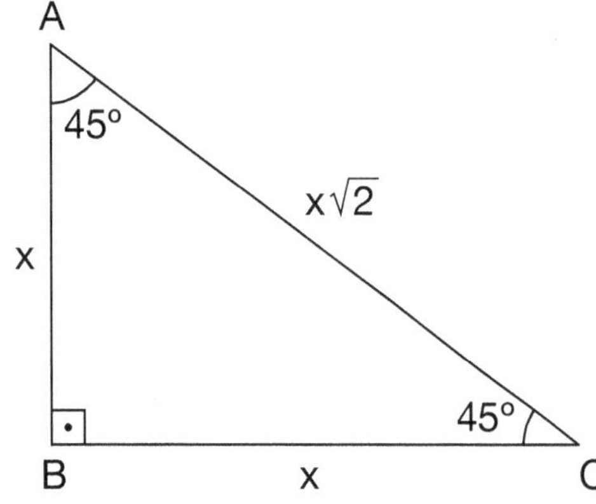

If $x\sqrt{2} = 3\sqrt{2}$

$x = 3$

Correct Answer : B

9. Solution:

If midpoint between A(6, 8) an B(a, b) is (10, 14), then

$\frac{6+a}{2} = 10$, $6 + a = 20$, $a = 14$

$\frac{8+b}{2} = 14$, $8 + b = 28$, $b = 20$

Coordinate point of B is (14, 20)

Correct Answer : B

10. Solution:

Coordinate points: (4, 5) and (7, 11)

slope = $m = \frac{y_2 - y_1}{x_2 - x_1} = \frac{11-5}{7-4} = \frac{6}{3} = 2$

Correct Answer : B

11. Solution:

Area of $\triangle ABC = \frac{base \times h}{2}$

$= \frac{\cancel{2}x(x+1)}{\cancel{2}}$

$= x(x + 1)$

$= x^2 + x$

Correct Answer : C

12. Solution:

A(4, –5) reflected over the y-axis is

$A^1(-4, -5)$

Correct Answer : D

13. Solution:

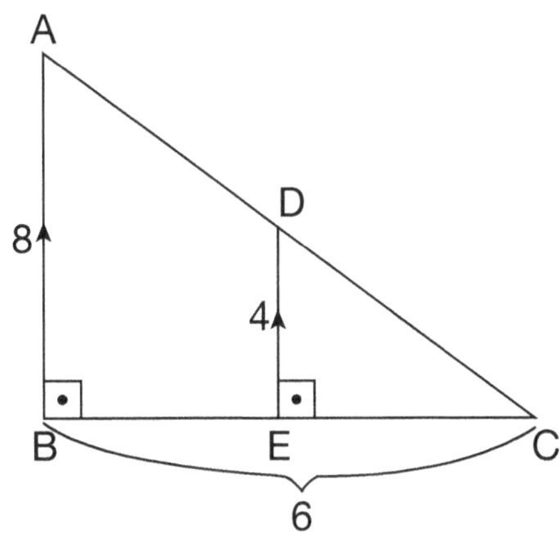

From pythagorean theorem:

$$|AC|^2 = 6^2 + 8^2 = 10^2$$

$$\sqrt{|AC|} = \sqrt{100}, \text{then} |AC| = 10$$

$$\frac{4}{8} = \frac{|DC|}{|AC|} \; , \; \frac{1}{2} = \frac{|DC|}{10}$$

$$|DC| = 5$$

$$\frac{1}{2} = \frac{|EC|}{|BC|} \; , \; \frac{1}{2} = \frac{|EC|}{6}$$

$$|EC| = 3$$

$$|DC| + |EC| = 5 + 3 = 8$$

Correct Answer : D

14. Solution:

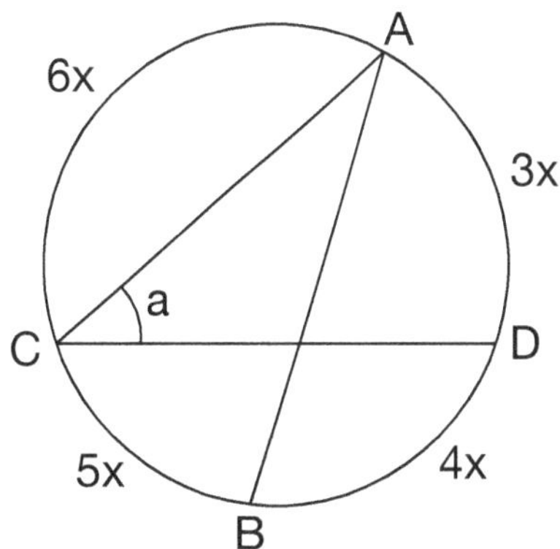

$6x + 5x + 4x + 3x = 360°$

$18x = 360°$

$x = 20 \quad , \quad 2a = 3x$

$2a = 60° \quad , \quad a = 30°, \angle ACD = 30°$

Correct Answer : C

15. Solution:

$3a + 5a = 96°$

$8a = 96° \longrightarrow a = 12° \longrightarrow 3a = 3 \cdot 12 = 36°$

Correct Answer : C

16. Solution:

From pythagorean theorem:

$8^2 + |BC|^2 = 17^2$, then $BC = 15$

$|CD| = 27 - 15 = 12$ then $|CE| = 13$

$\sin\alpha = \frac{12}{13}$

$\cot\beta = \frac{15}{8}$

$\sin\alpha . \cot\beta = \frac{12}{13} \cdot \frac{15}{8} = \frac{45}{26}$

Correct Answer : D

17. Solution:

$\text{Are of circle} = \pi r^2$

$\text{Are of square} = x^2$

$x^2 = \pi r^2$, then $x = r\sqrt{\pi}$

$\text{Perimeter of square} = 4r\sqrt{\pi}$

$\text{Circum frence of circle} = 2\pi r$

$$\frac{\text{Perimeter of square}}{\text{Circum frence of circle}} = \frac{4r\sqrt{\pi}}{2\pi r}$$

$$= \frac{2\sqrt{\pi}}{\pi}$$

Correct Answer : A

18. Solution:

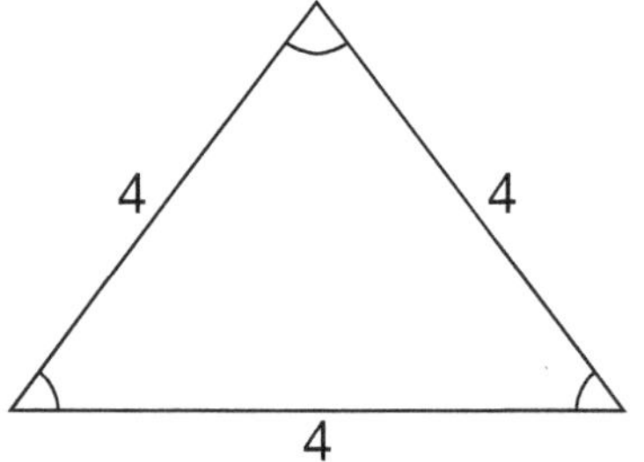

$\text{Area} = \frac{a^2\sqrt{3}}{4}$

$\text{Area} = \frac{(4)^2\sqrt{3}}{4}$

$= \frac{16\sqrt{3}}{4} = 4\sqrt{3}$

Correct Answer : C

American Math Academy

19. Solution:

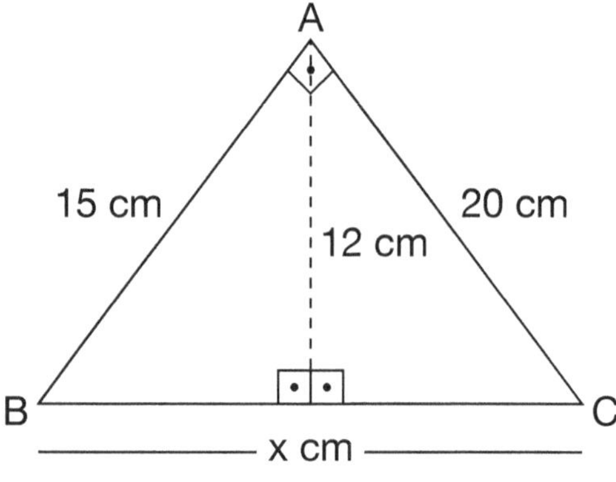

Rule: 12 cm · x = 15 cm · 20 cm

$x = \frac{15\,cm \cdot 20\,cm}{12\,cm} = \frac{300\,cm^2}{12\,cm} = 25\,cm$

x = 25 cm

Correct Answer : C

SAMPLE TEST SOLUTIONS

20. Solution:

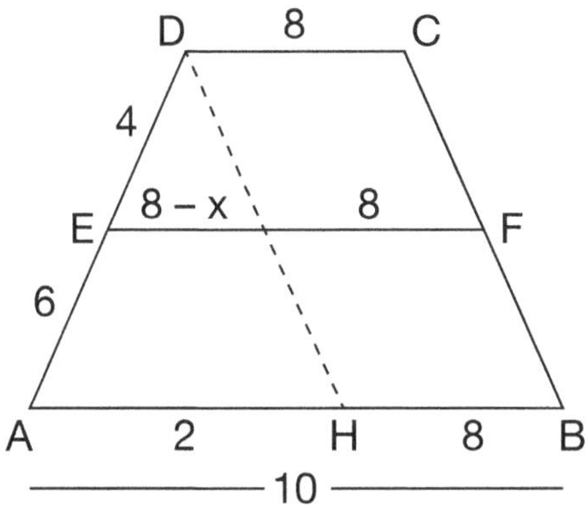

From similarity theorem

$\frac{x-8}{2}=\frac{4}{10}$

$10(x - 8) = 8$

$10x - 80 = 8$ $10x = 88$ $x = 8.8$

Correct Answer : C

21.

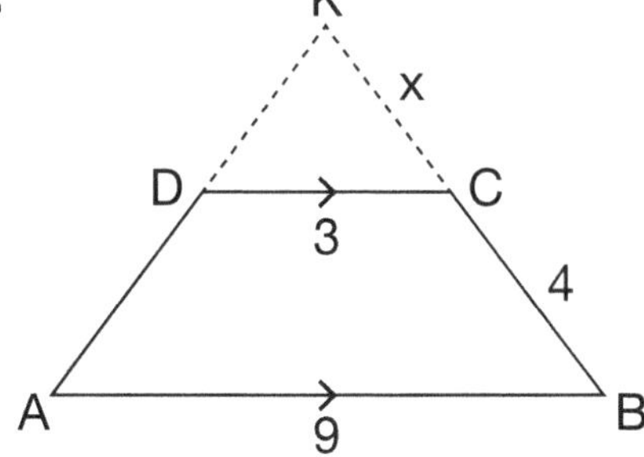

From similarity theorem:

$\frac{x}{x+4}=\frac{3}{9}$, $9x = 3x + 12$,

$9x - 3x = 12$ then $6x = 12$

$x = 2$

Correct Answer : A

22. Solution:

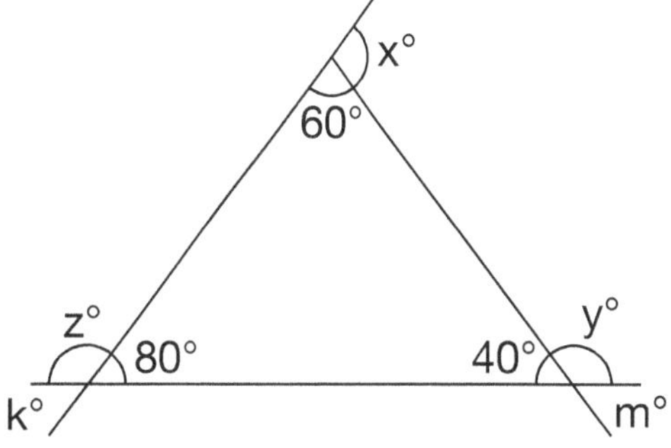

$x = 120°$

$y = 140°$

$z = 100°$

$k = 80°$

$m = 40°$

Correct Answer : B

23. Solution:

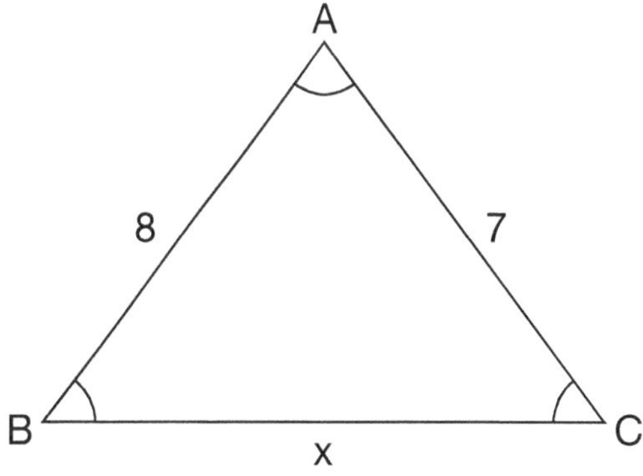

From triangle inequality theorem:

$8 - 7 < x < 8 + 7$

$1 < x < 15$

Correct Answer : D

24. Solution:

Area of $\triangle$ ABC $= \frac{b \cdot h}{2} = \frac{3\,cm \cdot 18\,cm}{2}$

$= \frac{54\,cm^2}{2} = 27\,cm^2$

Correct Answer : B

25. Solution:

$P = x + 1 + x - 2 + x + 2 = 3x + 1$

Correct Answer : C

26. Solution:

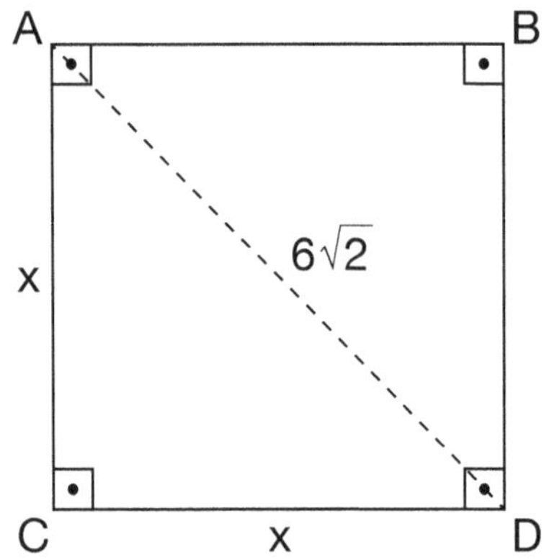

From pythagorean theorem:

$x^2 + x^2 = (6\sqrt{2})^2$

$2x^2 = 72$

$x^2 = 36, \quad x = 6$

$A(ABCD) = 6^2 = 36$

Correct Answer : C

27. Solution:

$4k + 5k + 9k = 180^o$

$18k = 180^o$

$k = 10^o$

smallest angle = 4k

$= 40^o$

Correct Answer : A

28. Solution:

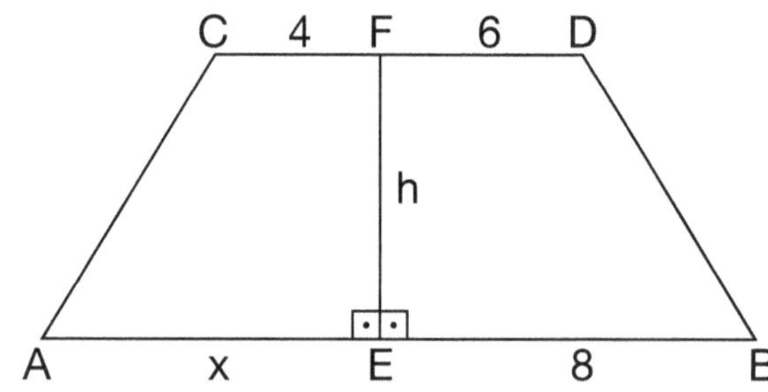

If A(AEFC) = A(EBDF)

$\frac{(6+8)\cancel{h}}{2} = \frac{(4+x)\cancel{h}}{2}$

$\frac{14}{\cancel{2}} = \frac{4+x}{\cancel{2}}$

$14 = 4 + x \qquad 10 = x$

Correct Answer : C

29. Solution:

$P = 2L + 2W$

$W = \frac{P - 2L}{2} = \frac{P}{2} - L$

Correct Answer : B

SAMPLE TEST SOLUTIONS

30. Solution:

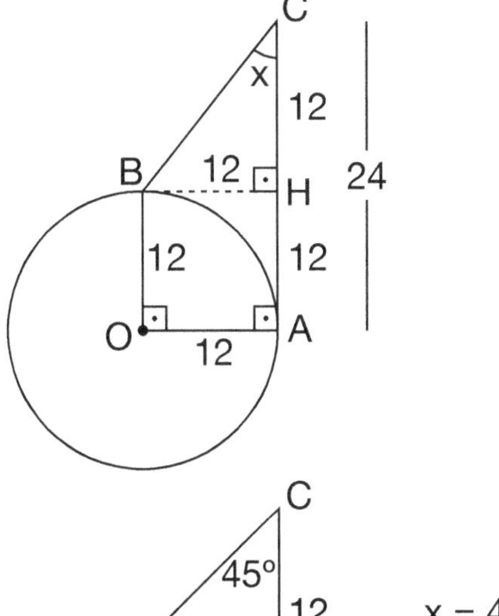

$x = 45^\circ$

Correct Answer : D

31. Solution:

From circle equation:

$(x - h)^2 + (y - k)^2 = r^2$

$(x - 3)^2 + (y - 6)^2 = 16$

Correct Answer : C

32. Solution:

Let $a = 2k, b = 3k, c = 6k$

$$\frac{1}{8k^3} + \frac{1}{27k^3} + \frac{1}{216k^3} = \frac{1}{6}$$

$$\frac{36}{216k^3} = \frac{1}{6} \quad , \quad 216k^3 = 216$$

$$k^3 = 1$$

$$k = 1$$

$V_c = 216k^3 = 216$

Correct Answer : A

33. Solution:

Suppose $r = 4 \quad V = \pi r^2$

$h = 5 \quad V = \pi \cdot 16 \cdot 5 = 80\pi$

$r \longrightarrow 25\%$ increase $r = 5$

$h \longrightarrow 20\%$ decrease $h = 4$

$V = \pi r^2 = 25 x 4 x \pi = 100\pi$

$\frac{100\pi - 80\pi}{80\pi} = \frac{1}{4} = 25\%$ increase

Correct Answer : C

34. Solution:

$\text{Cosine}\, B = \frac{\text{adjacent}}{\text{hypotenuse}} = \frac{3}{5}$

$\text{Sine}\, B = \frac{\text{opposite}}{\text{hypotenuse}} = \frac{4}{5}$

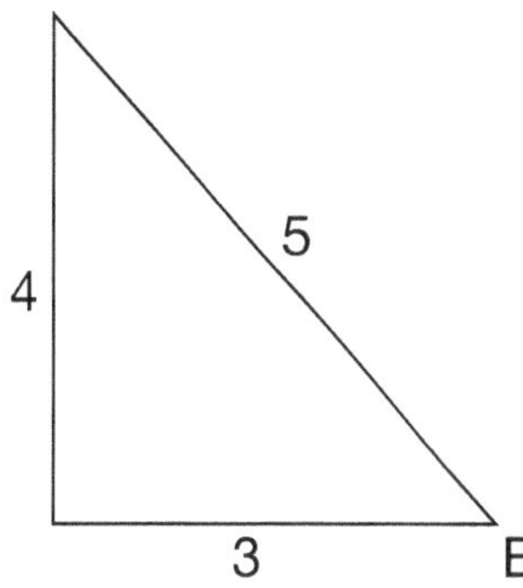

Ratio of the largest side to the shortest side

$= \frac{5}{3}$, or 5 to 3.

Correct Answer : D

LOGIC AND REASONING TEST SOLUTIONS

1. **Solution:**

   A proposition is a statement that is either true or false.

   Correct Answer: C

2. **Solution:**

   Negation is the opposite truth value of a given proposition.

   Correct Answer: A

3. **Solution:**

   The symbol $\wedge$ represents logical conjunction (AND).

   Correct Answer: B

4. **Solution:**

   The symbol $\longleftrightarrow$ represents a biconditional statement.

   Correct Answer: D

5. The statement "It is raining, or the ground is wet" is represented symbolically as $p \vee q$.

   Correct Answer: C

6. If a bird can sing, then it is not a penguin.

   Converse: If a bird is not a penguin, then it can sing.

   Inverse: If a bird cannot sing, then it is a penguin.

   Contrapositive: If a bird is a penguin, then it cannot sing.

7. If a car is electric, then it does not use gasoline.

   Converse: If a car does not use gasoline, then it is electric.

   Inverse: If a car is not electric, then it uses gasoline.

   Contrapositive: If a car uses gasoline, then it is not electric.

8. If a shape has four equal sides, then it is a square.

   Converse: If a shape is a square, then it has four equal sides.

   Inverse: If a shape does not have four equal sides, then it is not a square.

   Contrapositive: If a shape is not a square, then it does not have four equal sides.

9. If a planet orbits the sun, then it is not outside our solar system.

   Converse: If a planet is not outside our solar system, then it orbits the sun.

   Inverse: If a planet does not orbit the sun, then it is outside our solar system.

   Contrapositive: If a planet is outside our solar system, then it does not orbit the sun.

10. If a number is even, then it is not prime.

    Converse: If a number is not prime, then it is even.

    Inverse: If a number is odd, then it is prime.

    Contrapositive: If a number is prime, then it is odd.

    (Note: The only even prime number is 2. All other even numbers are not prime.)

ALGEBRAIC PROOFS TEST SOLUTIONS

1. **Solution:**

Algebraic Steps	**Properties**
$7x - 3(2x + 4) = 10$	Given
$7x - 6x - 12 = 10$	Distributive Property
$x - 12 = 10$	Combining Like Terms
$x = 22$	Addition and Subtraction

2. **Solution:**

Algebraic Steps:

1. $3x + 5(x - 3) = 17$ I Given
2. $3x + 5x - 15 = 17$ I Distributive Property
3. $8x - 15 = 17$ I Combining Like Terms
4. $8x = 32$ I Addition Property (Add 15 to both sides)
5. $x = \frac{32}{8}$ I Division Property
6. $x = 4$ I Simplify

3. **Solution:**

Algebraic Steps:

1. $2x + 12 = 3x + 18$ I Given
2. $2x - 3x = 18 - 12$ I Subtraction Property (subtract 3x from both sides and subtract 12 from both sides.)
3. $-x = 6$ I Combining Like Terms
4. $x = -6$ I Multiplication Property (multiply both sides by –1 to solve for x)

4. **Solution:**

Algebraic Steps:

1. $4x + 14 = 5x + 20$ I Given
2. $4x - 5x = 20 - 14$ I Subtraction Property (subtract 5x from both sides and subtract 14 from both sides.)
3. $-x = 6$ I Combining Like Terms
4. $x = -6$ I Multiplication Property (multiply both sides by –1 to solve for x)

GEOMETRIC PROOFS SOLUTIONS:

1. Solution:

Statements	Reasons
1. B is the midpoint of $\overline{AC}$	1. Given
2. AB = BC	2. Definition of midpoint
3. C is the midpoint of $\overline{BD}$	3. Given
4. BC = CD	4. Definition of midpoint
5. AB + BC = BC + CD	5. Addition Property
6. AB + BC = AC, BC + CD = BD	6. Segment Addition Postulate
7. AC = BD	7. Substitution

2. Solution:

Statements	Reasons
1. $\angle 2$ and $\angle 3$ are supplementary.	1. Linear Pair Postulate
2. $m\angle 2 + m\angle 3 = 180°$	2. Substitution Property of Equality
3. $\angle 1$ and $\angle 2$ are a linear pair.	3. Given
4. $\angle 1$ and $\angle 2$ are supplementary.	4. Definition of Congruent Angles
5. $m\angle 1 + m\angle 2 = 180°$	5. Converse of the Corresponding Angles Postulate
6. $m\angle 1 = m\angle 3$	6. Definition of Supplementary Angles
7. $\angle 1 \cong \angle 3$	7. Definition of a Linear Pair
8. A || B	8. Definition of Supplementary Angles

3. Solution:

Statements	Reasons
1. a || b, c ⊥ a	1. Given
2. $m\angle 2 = 90°$	2. Definition of perpendicular lines
3. $\angle 2 \cong \angle 6$	3. Corresponding Angles Theorem
4. $m\angle 2 = m\angle 6$	4. Definition of congruent angles
5. $m\angle 6 = 90°$	5. Transitive Property of Equality
6. c ⊥ b	6. Definition of perpendicular lines

GEOMETRIC PROOFS SOLUTIONS:

4. Solution:

Statements	Reasons
1. $\angle 3$ and $\angle 4$ are vertical $\angle$ s	1. Given
2. $m\angle 3 + m\angle 1 = 180$ $m\angle 1 + m\angle 4 = 180$	2. Angle Addition Postulate
3. $m\angle 3 + m\angle 1 = m\angle 1 + m\angle 4$	3. Substitution
4. $m\angle 1 = m\angle 1$	4. Reflexive Property
5. $m\angle 3 = m\angle 4$	5. Subtraction Property
6. $\angle 3 \cong \angle 4$	6. Definition of $\cong$ Angles

5. Solution:

Statements	Reasons
$m\angle A + m\angle B$ is the measure of a right angle.	By definition of a right angle.
A right angle measures 90°.	Definition of a right angle.
Therefore, $m\angle A + m\angle B = 90°$	Transitive Property.

6. Solution:

Statements	Reasons
$m\angle A + m\angle B$ is the measure of a straight angle on line segment AB.	By definition of straight angle.
A straight angle measures 180°.	Definition of a straight angle.
Therefore, $m\angle A + m\angle B = 180°$	Transitive Property.

ANGLE RELATIONSHIPS TEST SOLUTIONS

1. **Solution:**

$2x + 5^\circ + 90^\circ + 3x - 10^\circ = 180^\circ$

$5x + 85^\circ = 180^\circ$

$\quad -85^\circ \quad -85$

$+$ ________________

$5x = 95^\circ,\ x = 19^\circ$

Correct Answer : D

2. **Solution:**

$5x + 4x + 27^\circ = 180^\circ$

$9x + 27^\circ = 180^\circ$

$\quad -27^\circ \quad -27^\circ$

$+$ ________________

$9x = 153^\circ,\ x = 17^\circ$

Correct Answer : B

3. **Solution:**

$88^\circ + x + 44^\circ = 180^\circ$

$132^\circ + x = 180^\circ$

$-132^\circ \quad -132^\circ$

$+$ ________________

$x = 48^\circ$

Correct Answer : D

4. **Solution:**

$3x - 10^\circ + 2x + 70^\circ = 180^\circ$

$5x + 60^\circ = 180^\circ$

$\quad -60^\circ \quad -60^\circ$

$+$ ________________

$5x = 120^\circ,\ x = 24^\circ$

Correct Answer : C

5. **Solution:**

$30^\circ + 3x + 15^\circ = 90^\circ$

$45^\circ + 3x = 90^\circ$

$-45^\circ \quad -45^\circ$

$+$ ________________

$3x = 45^\circ,\ x = 15^\circ$

Correct Answer : A

6. **Solution:**

Since m // n, then

$3x + 31^\circ = 5x - 27^\circ$

$-3x \quad -3x$

$+$ ________________

$31^\circ = 2x - 27^\circ$

$27^\circ \quad +27^\circ$

$+$ ________________

$58^\circ = 2x$

$29^\circ = x$

Correct Answer : B

7. **Solution:**

$130^\circ + 40^\circ + 90^\circ + 5x = 360^\circ$

$260^\circ + 5x = 360^\circ$

$-260^\circ \quad -260^\circ$

$+$ ________________

$5x = 100^\circ$

$x = 20^\circ$

Correct Answer : C

ANGLE RELATIONSHIPS TEST SOLUTIONS

8. Solution:

$135^\circ = 3x$

$\frac{135^\circ}{3} = x$

$45^\circ = x$

Correct Answer : D

9. Solution:

Since AB // DE

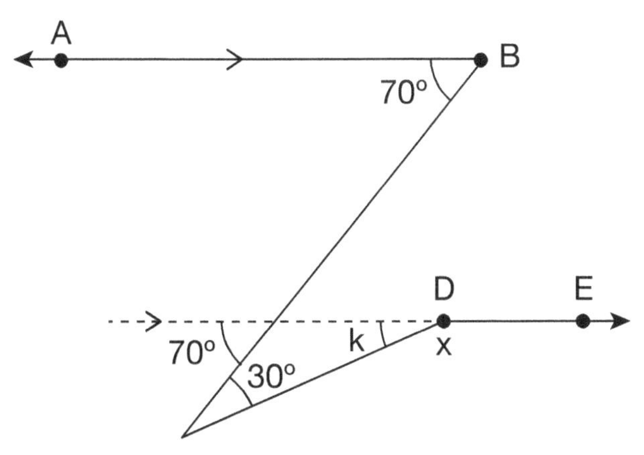

$k + 30^\circ = 70^\circ$

$k = 40^\circ$

$x + k = 180^\circ$

$x + 40^\circ = 180^\circ$

$x = 140^\circ$

Correct Answer : B

10. Solution:

$5(x + 30)^\circ + 3(x + 2)^\circ = 180^\circ$

$5x + 150^\circ + 3x + 6^\circ = 180^\circ$

$$\begin{array}{rl} & 8x + 156^\circ = 180^\circ \\ + & \quad -156^\circ \quad -156 \\ \hline & 8x = 24^\circ,\ x = 3^\circ \end{array}$$

Correct Answer : A

11. Solution:

$3x + x = 90^\circ,\ 4x = 90^\circ$

$x = \frac{90^\circ}{4} = \frac{45^\circ}{2} = 22.5^\circ$

$5y + 15^\circ = 90^\circ,\ 5y = 75^\circ$

$y = 15^\circ$

$x + y = 22.5 + 15^\circ$

$x + y = 37.5^\circ$

Correct Answer : C

12. Solution:

Since $\angle AOD = 90^\circ$

$$\begin{array}{rl} & 2x + 5^\circ + 3x + 10^\circ = 90^\circ \\ & 5x + 15^\circ = 90^\circ \\ + & \quad -15^\circ \quad -15^\circ \\ \hline & 5x = 75^\circ\ ,\ x = 15^\circ \end{array}$$

Correct Answer : B

INTERIOR AND EXTERIOR ANGLES TEST SOLUTIONS

1. **Solution:**

   Since $\angle 1$ and $\angle 6$ are alternate exterior angles, then $\angle 1 \cong \angle 6$

   So, m $\angle$ 1 = 70°

2. **Solution:**

   Since $\angle 7$ and $\angle 3$ are corrosponding, then $\angle 7 \cong \angle 3$

   So, m $\angle$ 3 = 110°

3. **Solution:**

   Alternate interior angles:

   $\angle 4 \cong \angle 7$ and $\angle 2 \cong \angle 5$

4. **Solution:**

   Alternate exterior angles:

   $\angle 1$, $\angle 6$ and $\angle 3$, $\angle 8$

5. **Solution:**

   Corresponding angles:

   $\angle 1 \cong \angle 5$, $\angle 3 \cong \angle 7$, $\angle 4 \cong \angle 8$ and $\angle 2 \cong \angle 6$

6. **Solution:**

   Supplementary angles:

   $\angle 1$ and $\angle 3$, $\angle 2$ and $\angle 4$, $\angle 5$ and $\angle 7$

   $\angle 6$ and $\angle 8$

7. **Solution:**

   Since m $\angle$ 5 = 75° and m $\angle 5$ + m $\angle$ 7 = 180°

   75° + m $\angle$ 7 = 180°

   m $\angle$ 7 = 105°

8. **Solution:**

   Since m $\angle$ 8 = 125° and

   m $\angle 4$ and m $\angle 8$ are corresponding

   angles, then m $\angle$ 4 = 125°

SEGMENT ADDITION POSTULATE TEST SOLUTIONS

1. Solution:

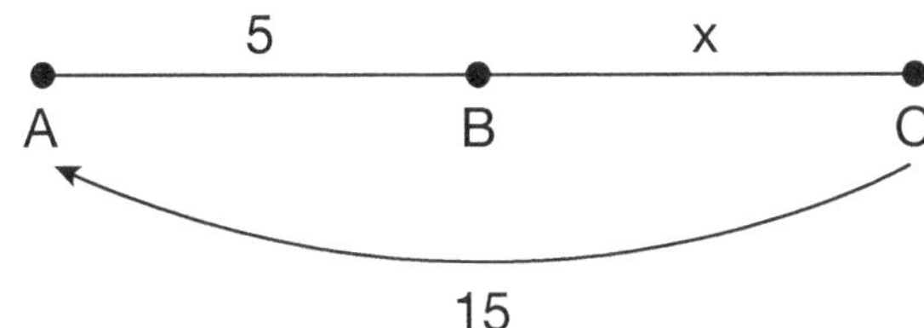

$5 + x = 15$

$x = 15 - 5$

$x = 10$

Correct Answer : B

2. Solution:

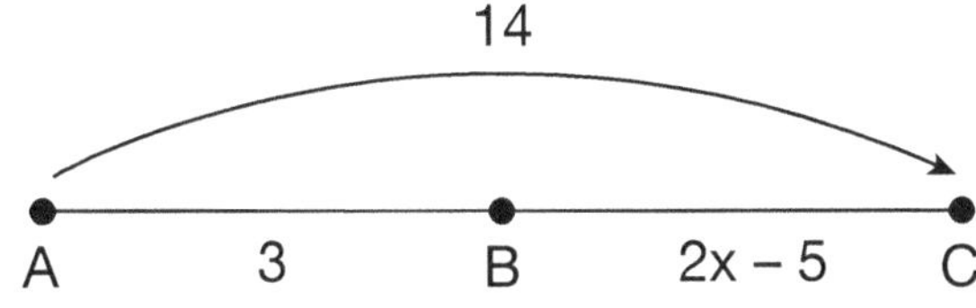

$3 + 2x - 5 = 14$

$2x - 2 = 14$

$+\quad +2 \quad +2$

$2x = 16$

$x = 8$

Correct Answer : B

3. Solution:

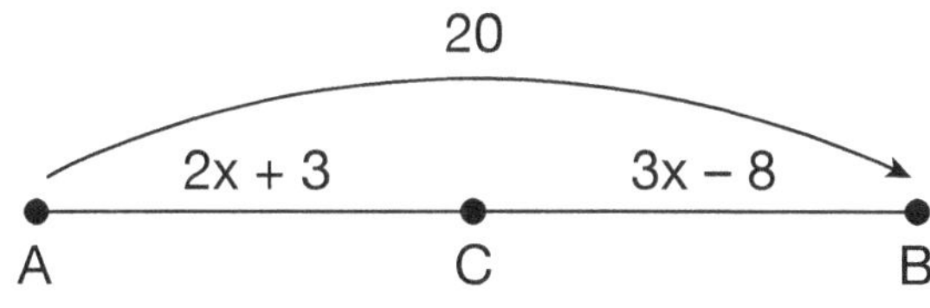

$2x + 3 + 3x - 8 = 20$

$5x - 5 = 20$

$+5 \quad +5$

$5x = 25, x = 5$

Correct Answer : A

American Math Academy

4. Solution:

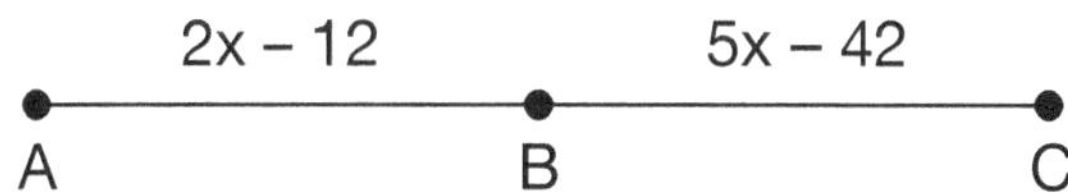

Since B is the midpoint of AC, then AB = BC

$2x - 12 = 5x - 42$

$+\quad -2x \quad -2x$

$-12 = 3x - 42$

$+\quad 42 \quad +42$

$30 = 3x$

$10 = x$

Correct Answer : B

5. Solution:

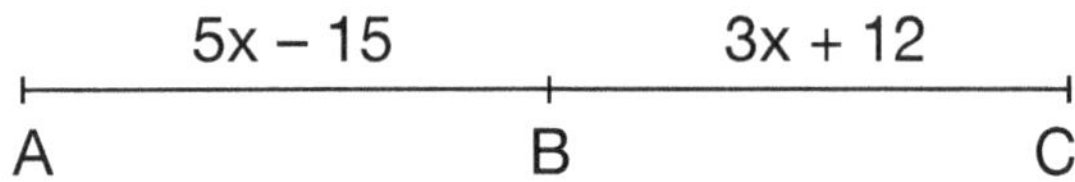

$5x - 15 = 3x + 12$

$2x = 27$

$x = 13.5$

Correct Answer : D

6. Solution:

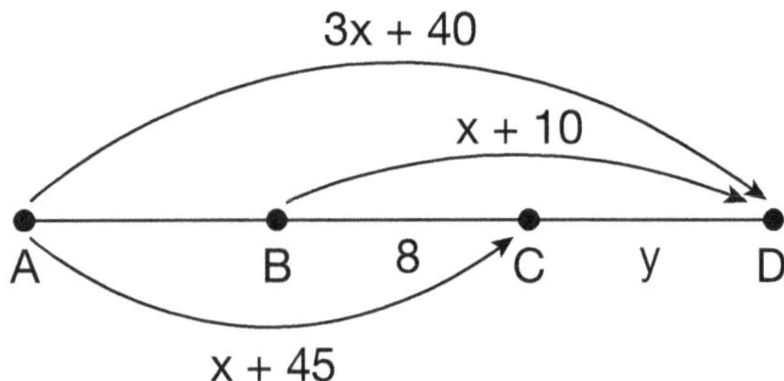

Since $BD = x + 10$ and $BC + CD = BD$

$8 + y = x + 10$

$y = x + 2$

$AD = AC + CD$

$3x + 40 = x + 45 + y$

$2x = 5 + y$, since $y = x + 2$

$2x = 5 + x + 2$, $x = 7$

Correct Answer : B

American Math Academy

ANGLE BISECTORS TEST SOLUTIONS

1. Solution:

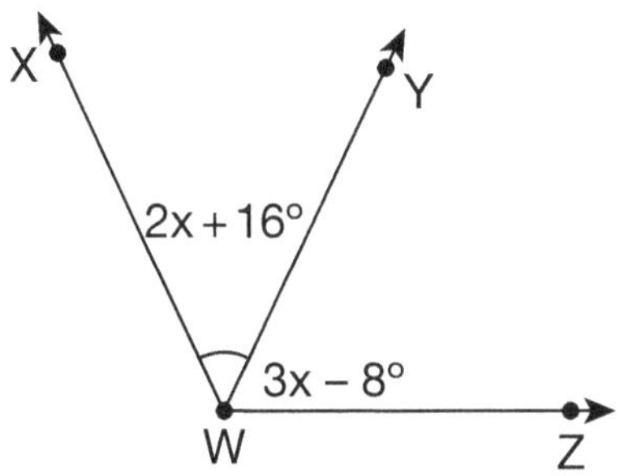

Since WY is bisector of

m ∠ XWY = m ∠ YWZ

$$\begin{aligned} 2x + 16° &= 3x - 8° \\ -2x &\quad -2x \\ \hline 16° &= x - 8° \\ 24° &= x \end{aligned}$$

Correct Answer : C

2. Solution:

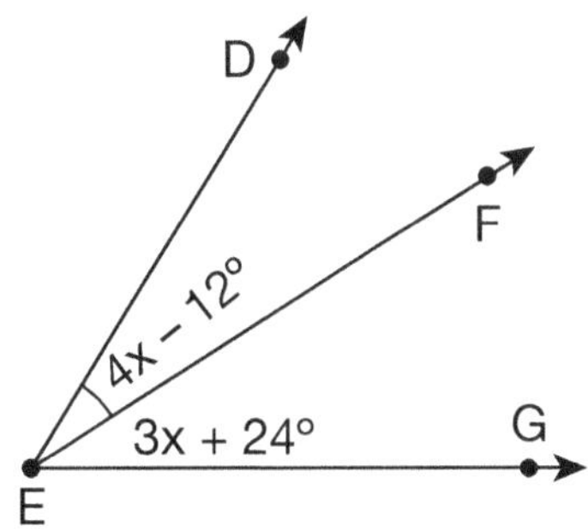

Since EF is bisector of m ∠ DEG,

then m ∠ DEF = m ∠ FEG,

$$\begin{aligned} 4x - 12° &= 3x + 24° \\ + \quad -3x &\quad -3x \\ \hline x - 12° &= 24° \\ x &= 36° \end{aligned}$$

Correct Answer : D

3. Solution:

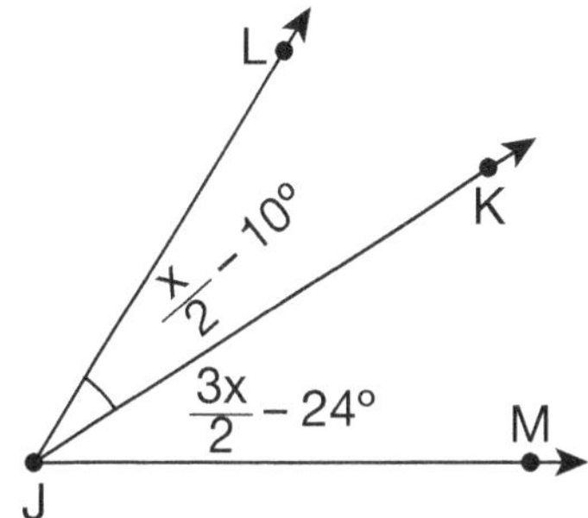

Since JK is bisector of m ∠ LJM,

then m ∠ LJK = m ∠ KJM

$$\begin{aligned} \frac{x}{2} - 10° &= \frac{3x}{2} - 24° \\ + \quad \frac{-x}{2} &\quad \frac{-x}{2} \\ \hline -10° &= \frac{3x}{2} - \frac{x}{2} - 24° \end{aligned}$$

$-10 = x - 24°$ $\qquad$ $14° = x$

Correct Answer : A

4. Solution:

$$\begin{aligned} 2x - 30 &= \frac{x-2}{3} + 15 \\ + \quad +30 &\quad +30 \\ \hline \end{aligned}$$

$2x = \dfrac{x-2}{3} + 45$ $\qquad$ $2x - \dfrac{x-2}{3} = 45$

$\dfrac{6x - x + 2}{3} = 45,$ $\qquad$ $\dfrac{5x+2}{3} = 45$

$5x + 2 = 135$ $\qquad$ $5x = 133$

$x = \dfrac{133}{5} = 26{\cdot}6$

m

A $\quad 2x - 30 \quad$ O $\quad \dfrac{x-2}{3} + 5 \quad$ B

Correct Answer : C

CLASSIFYING TRIANGLES TEST SOLUTIONS

1. Solution:

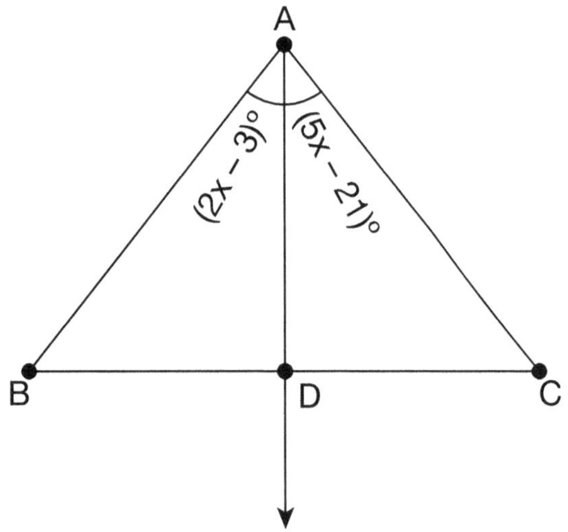

Since $\overrightarrow{AD}$ is bisects $\angle$ BAC then,

m $\angle$ BAD = $\angle$ DAC

$2x - 3^o = 5x - 21^o$

$+ \; -2x \qquad -2x$

$-3 = 3x - 21^o$

$+ \; 21 \qquad +21$

$18^o = 3x$

$6^o = x$

Correct Answer : B

2. Solution:

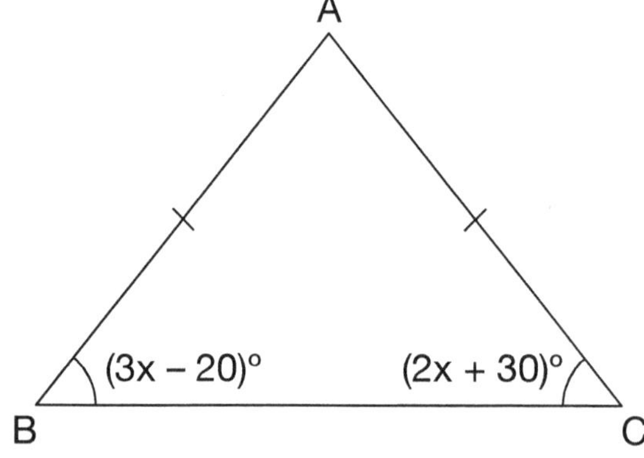

Since $\triangle$ BAC is an isosceles triangle

m $\angle$ ABC = m $\angle$ ACB

$3x - 20^o = 2x + 30^o$

$+ \; -2x \qquad -2x$

$x - 20^o = 30^o$

$x = 50^o$

Correct Answer : D

3. Solution:

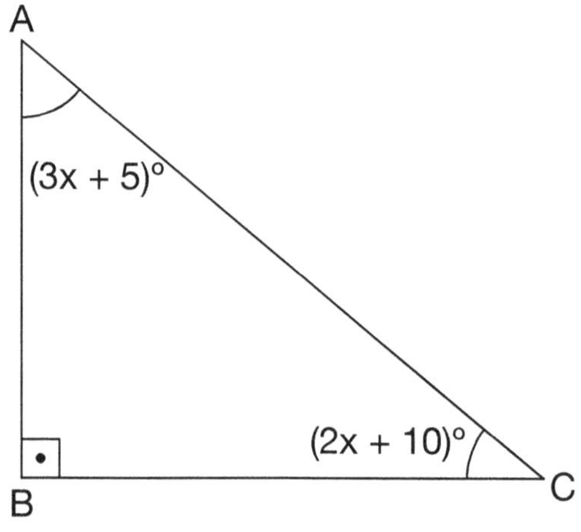

$3x + 5^o + 2x + 10^o = 90^o$

$5x + 15^o = 90^o$

$5x = 75^o \qquad x = 15^o$

Correct Answer : C

CLASSIFYING TRIANGLES TEST SOLUTIONS

4. Solution:

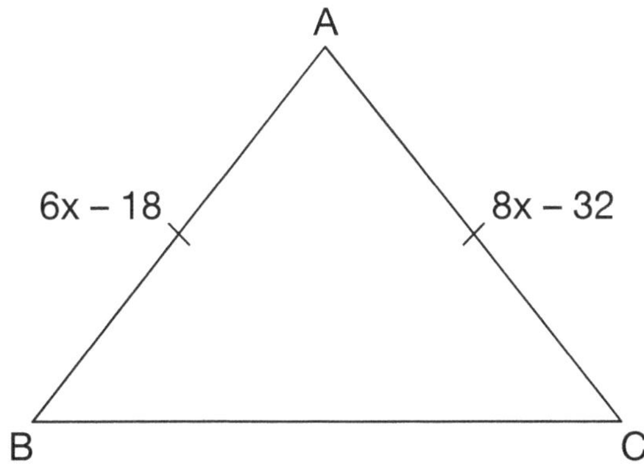

Since △ BAC an isosceles triangle and lengh of AB = AC, then

$$6x - 18 = 8x - 32$$
$$\underline{+\;-6x \qquad -6x}$$
$$-18 = 2x - 32$$
$$\underline{+\;32 \qquad +32}$$
$$14 = 2x$$
$$7 = x$$

Correct Answer : A

American Math Academy

5. Solution:

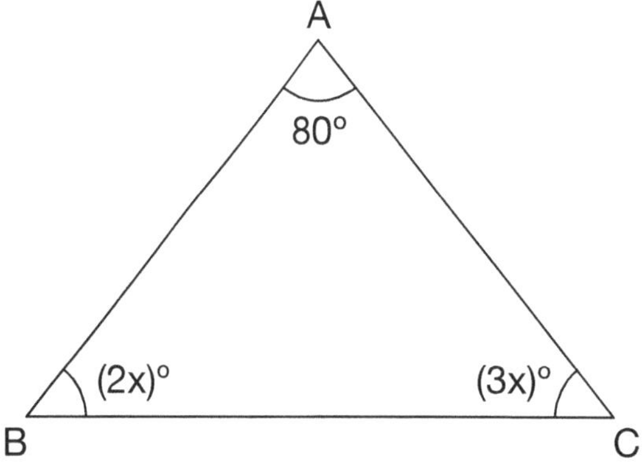

Since $m\angle A + m\angle B + m\angle C = 180^\circ$

$2x + 3x + 80^\circ = 180$

$5x + 80^\circ = 180^\circ$

$5x = 100^\circ$

$x = 20^\circ$

Correct Answer : D

INTERIOR AND EXTERIOR TRIANGLES TEST SOLUTIONS

1. **Solution:**

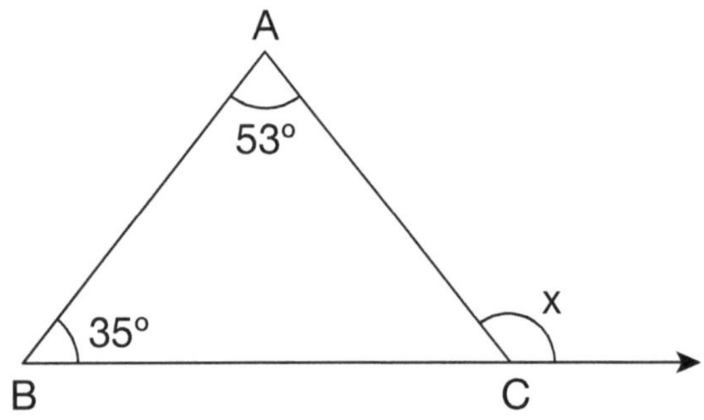

Since the exterior angle is equal to the sum of the two opposite interior angles,

$x = 35° + 53°$

$x = 88°$

Correct Answer : C

2. **Solution:**

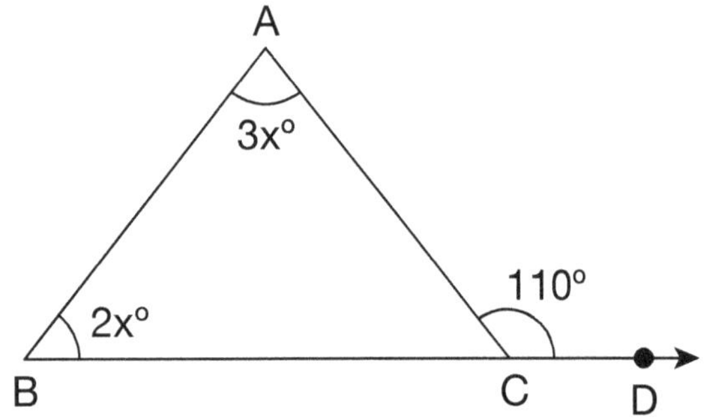

Since the exterior angle is equal to the sum of the two opposite interior angles,

$2x + 3x = 110°$

$5x = 110°$

$x = 22°$

Correct Answer : A

3. **Solution:**

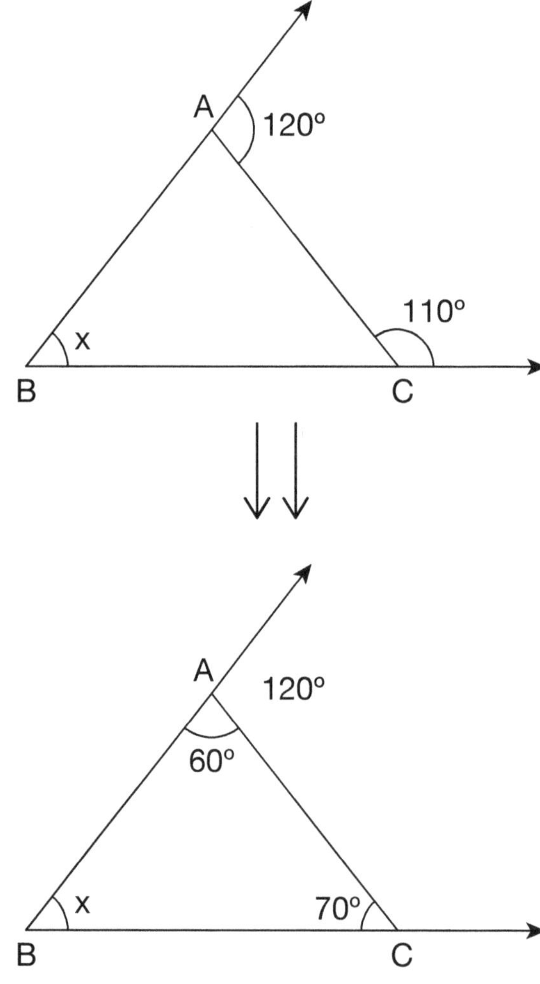

$x + 60° + 70° = 180$

$x + 130° = 180°$

$x = 50°$

Correct Answer : B

INTERIOR AND EXTERIOR TRIANGLES TEST SOLUTIONS

4. Solution:

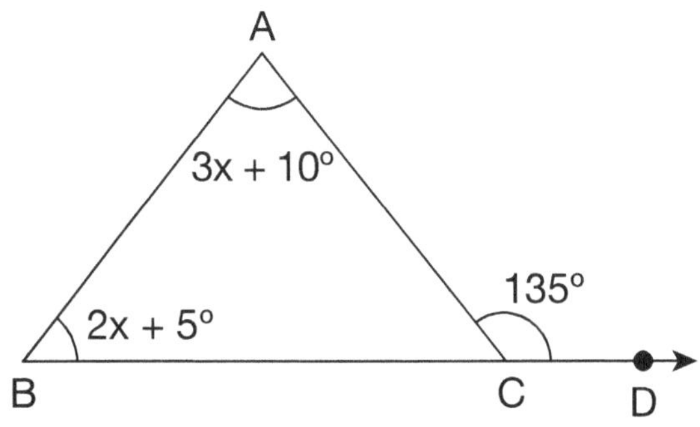

$3x + 10^\circ + 2x + 5^\circ = 135^\circ$

$5x + 15^\circ = 135^\circ$

$+ \quad -15^\circ \quad -15$

$5x = 120^\circ$

$x = 24^\circ$

Correct Answer : D

5. Solution:

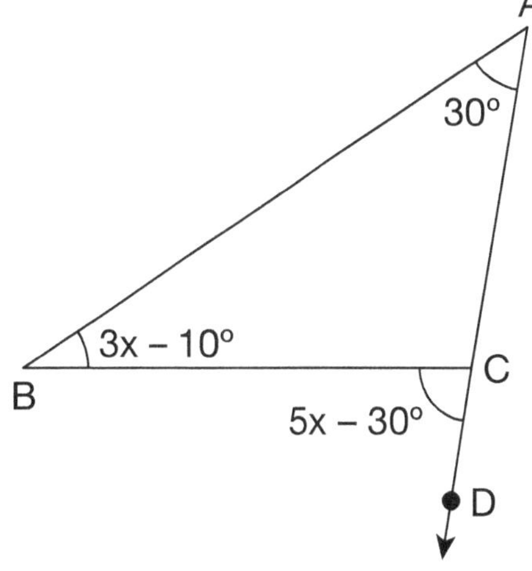

$3x - 10 + 30^\circ = 5x - 30^\circ$

$3x + 20^\circ = 5x - 30^\circ$

$+ \quad -3x^\circ \quad -3x$

$+20 = 2x - 30^\circ$

$+30^\circ \quad +30^\circ$

$50^\circ = 2x$

$25^\circ = x$

Correct Answer : B

TRIANGLE INEQUALITIES TEST SOLUTIONS

1. **Solution:**

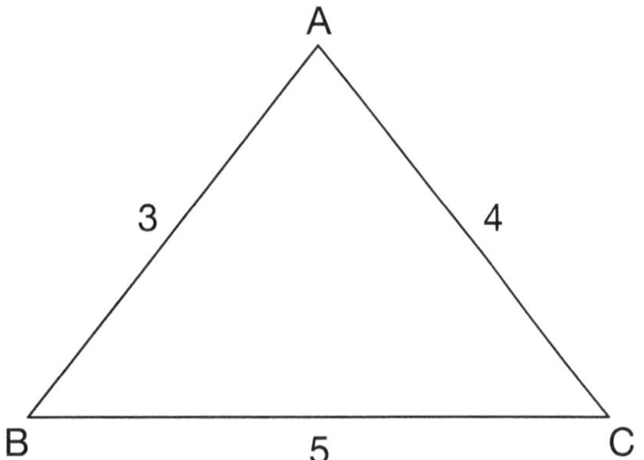

The shortest side is always opposite the smallest interior angle.

m $\angle$ C is the smallest angle

Correct Answer : C

2. **Solution:**

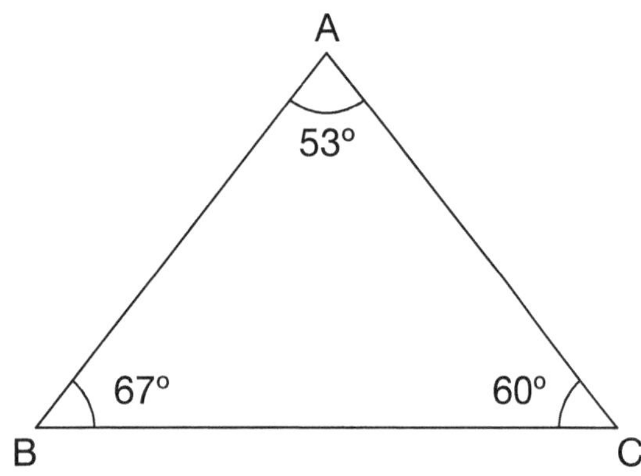

The shortest side is always opposite the smallest interior angle.

sides of shortest to longest:

$\overline{BC} < \overline{AB} < \overline{AC}$

Correct Answer : C

3. **Solution:**

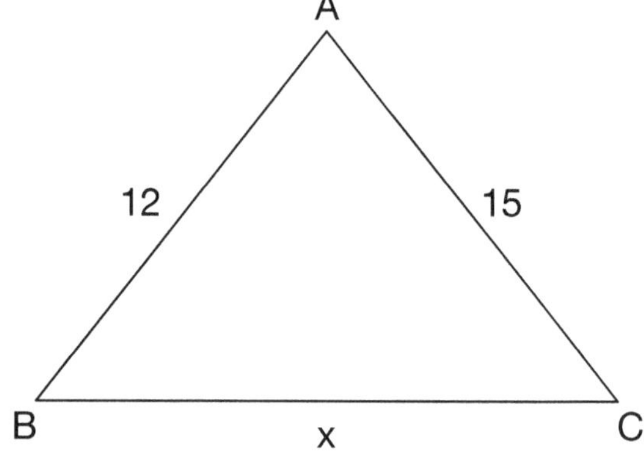

From triangle inequality theorem:

15 – 12 $\angle$ x $\angle$ 15 + 12

3 $\angle$ x $\angle$ 27

Correct Answer : D

TRIANGLE INEQUALITIES TEST SOLUTIONS

4. Solution:

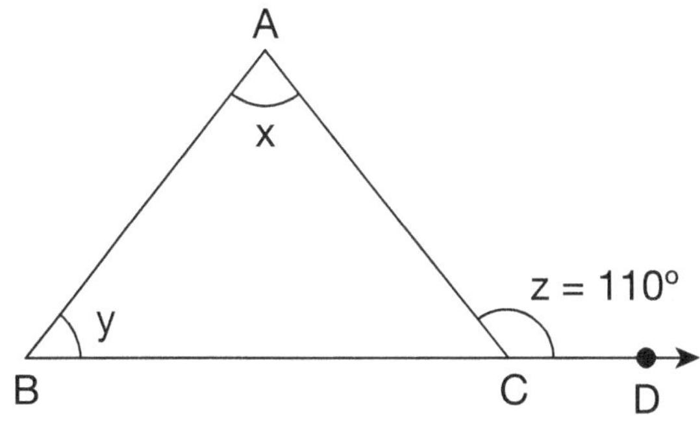

If $\angle y = \angle z - 50°$ then

$\angle y = 110° - 50°$

$\angle y = 60°$

$\angle x = 50°$

$\angle ACB = 70°$

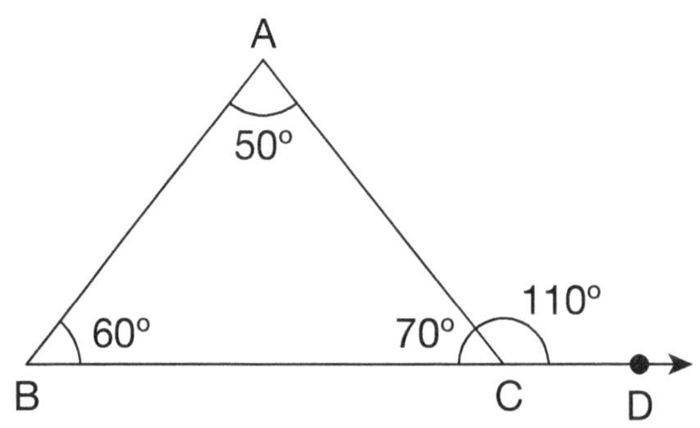

The greatest side is $\overline{AB}$

Correct Answer : A

5. Solution:

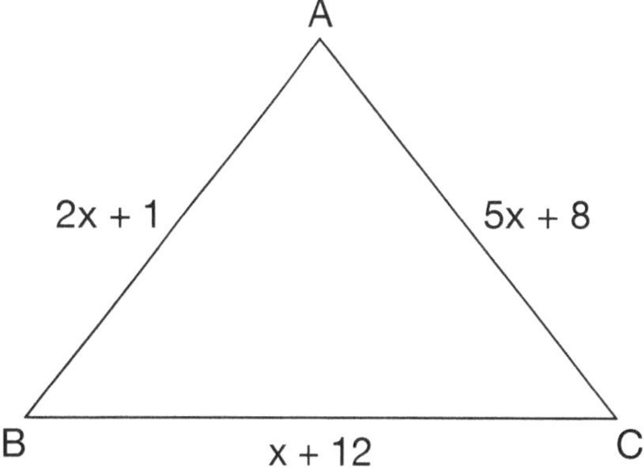

From triangle inequality theorem:

$2x + 1 + 5x + 8 > x + 12$ (1)

$5x + 8 + x + 12 > 2x + 12$ (2)

$2x + 1 + x + 12 > 5x + 8$ (3)

(1) $7x + 9 > x + 12$, then $6x > 3$, $x > \frac{1}{2}$

(2) $6x + 20 > 2x + 1$, then $2x > -19$, $x > \frac{-19}{2}$

(3) $3x + 13 > 5x + 8$, then $13 - 8 > 5x - 3x$

$5 > 2x$

$\frac{5}{2} > x$

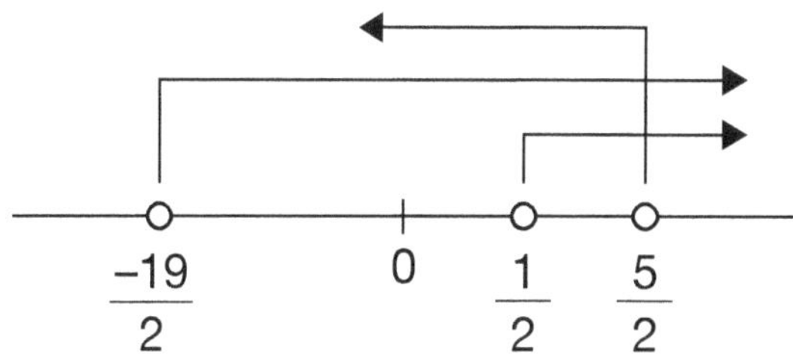

since x is betwen $\frac{3}{7}$ to $\frac{5}{2}$ from all options x can be only C

Correct Answer : C

SPECIAL RIGHT TRIANGLES TEST SOLUTIONS

1. Solution:

Rule:

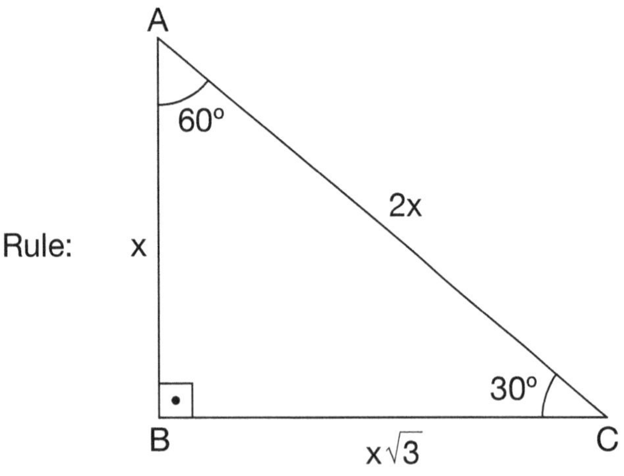

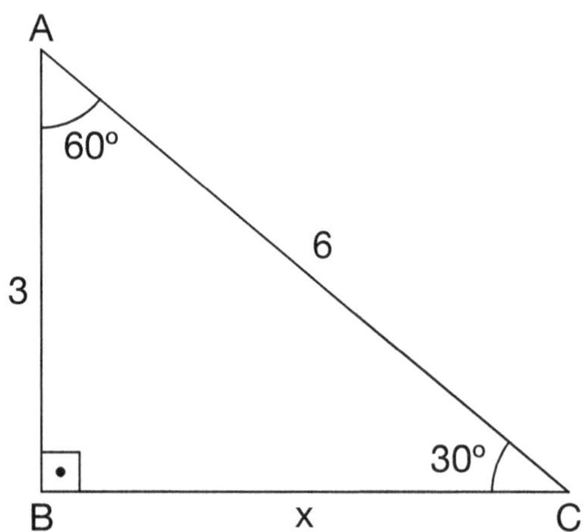

If 30° is equal to 3, then 60° is $3\sqrt{3}$

$x = 3\sqrt{3}$

Correct Answer : C

2. Solution:

Rule:

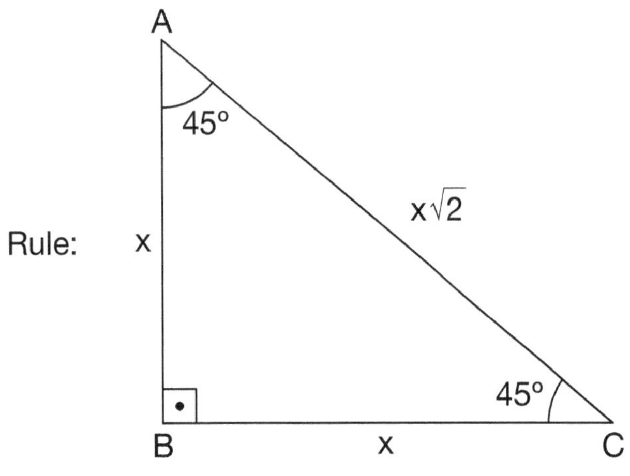

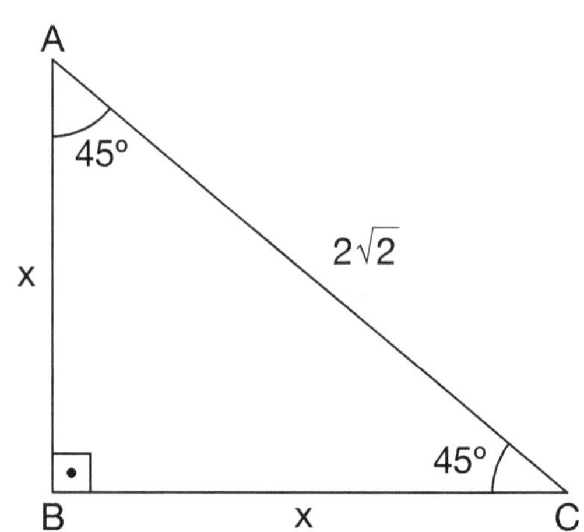

If $x\sqrt{2} = 2\sqrt{2}$, then $x = 2$

Correct Answer : A

SPECIAL RIGHT TRIANGLES TEST SOLUTIONS

3. Solution:

Rule:

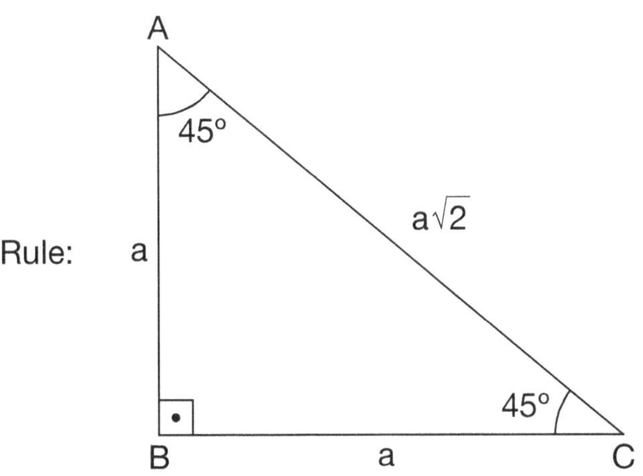

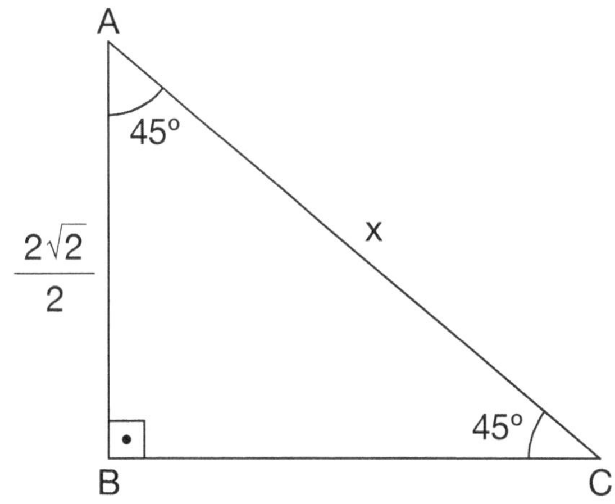

If $a = \sqrt{2}$, then $x = a\sqrt{2} = \sqrt{2} \times \sqrt{2} = 2$

Correct Answer : B

4. Solution:

Rule:

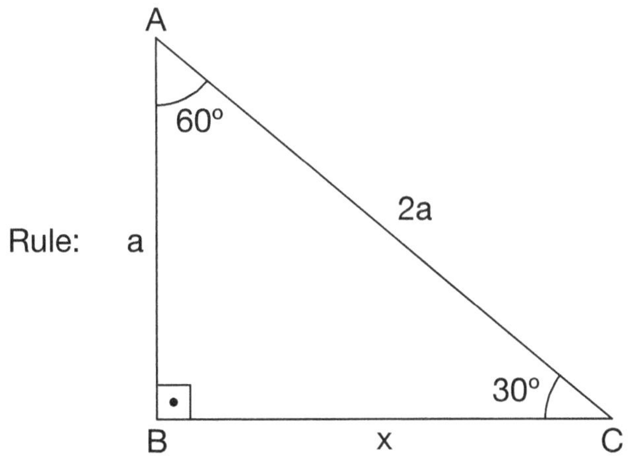

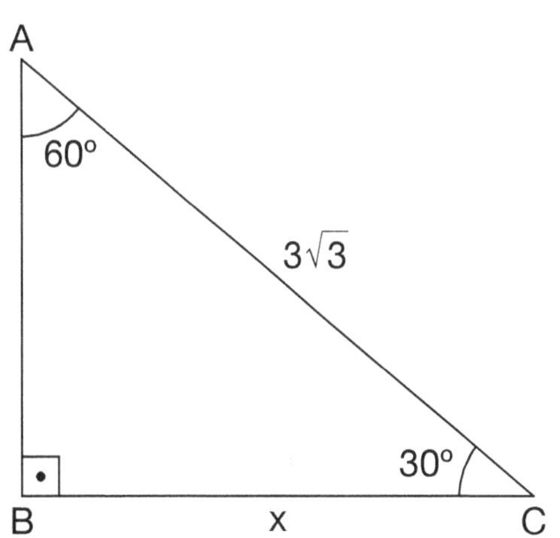

If $2a = 3\sqrt{3}$, then $a = \dfrac{3\sqrt{3}}{2}$

since $x = a\sqrt{3}$ then $\dfrac{3\sqrt{3}}{2} \times (\sqrt{3}) = \dfrac{3\sqrt{9}}{2} = \dfrac{3x3}{2} = \dfrac{9}{2}$

Correct Answer : D

5. Solution:

Rule:

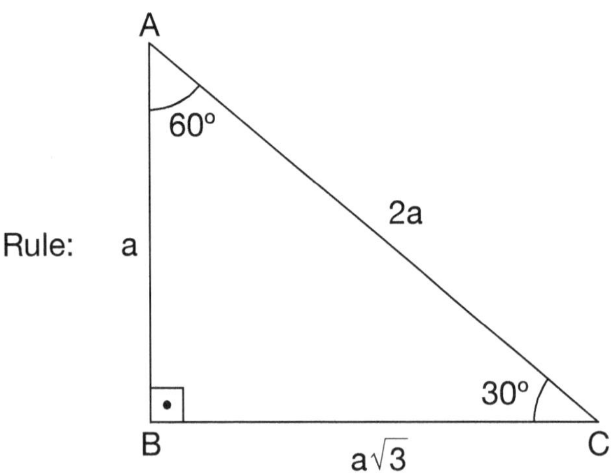

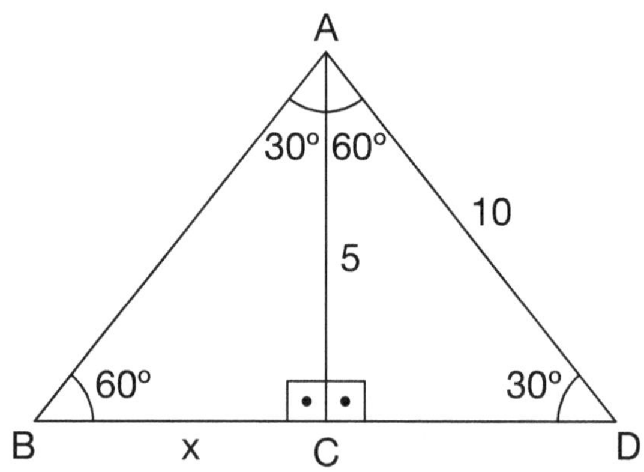

From △ ABC if AC = 5, then

$a\sqrt{3} = 5,\ a = \frac{5}{\sqrt{3}} \frac{\times \sqrt{3}}{\times \sqrt{3}} = \frac{5\sqrt{3}}{3}$

Since $x = a = \frac{5\sqrt{3}}{3}$

Correct Answer : C

PYTHAGOREAN THEOREM TEST SOLUTIONS

1. Solution:

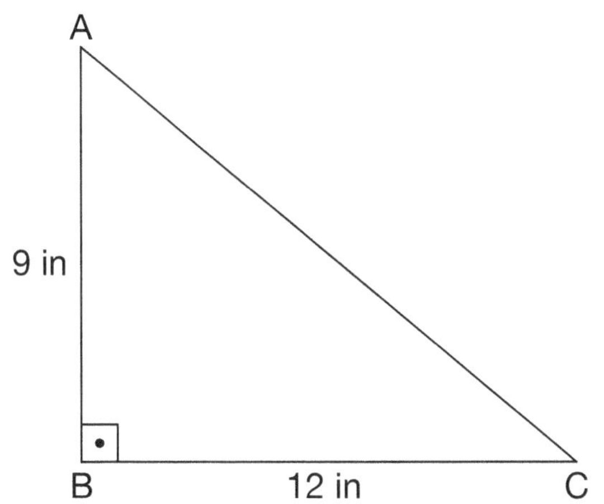

From pythagorean theorem $|AC|^2 = 9^2 + 12^2$

$|AC|^2 = 81\ in^2 + 144\ in^2$

$\sqrt{|AC|^2} = \sqrt{225\,in^2}$

AC = 15 in

Correct Answer : D

2. Solution:

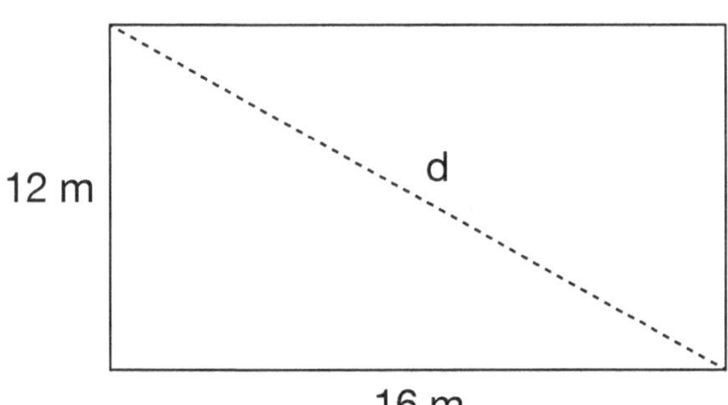

From pythagorean theorem:

$d^2 = (12\ m)^2 + (16\ m)^2$

$d^2 = 144\ m^2 + 256\ m^2$

$\sqrt{d^2} = \sqrt{400\,m^2}$

d = 20 m

Correct Answer : D

American Math Academy

3. Solution:

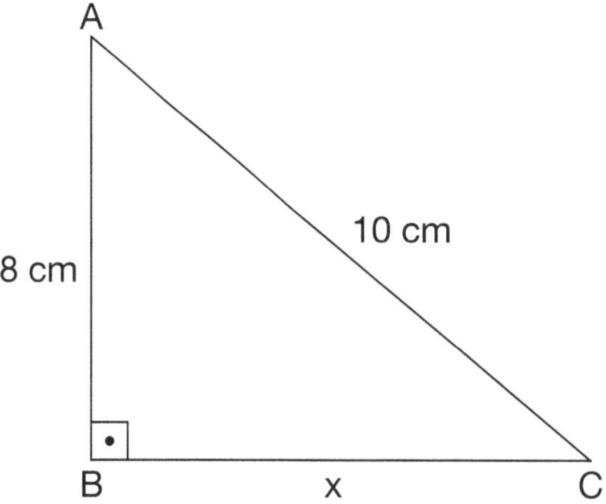

From pythagorean theorem;

$|AC|^2 = |AB|^2 + |BC|^2$

$(10\ cm)^2 = x^2 + (8\ cm)^2$

$100\ cm^2 = x^2 + 64\ cm^2$

$$\begin{array}{rcl} -64\ cm^2 & & -64\ cm^2 \\ + \\ \hline \end{array}$$

$\sqrt{36\,cm^2} = \sqrt{x^2}$

6 cm = x

Correct Answer : A

4. Solution:

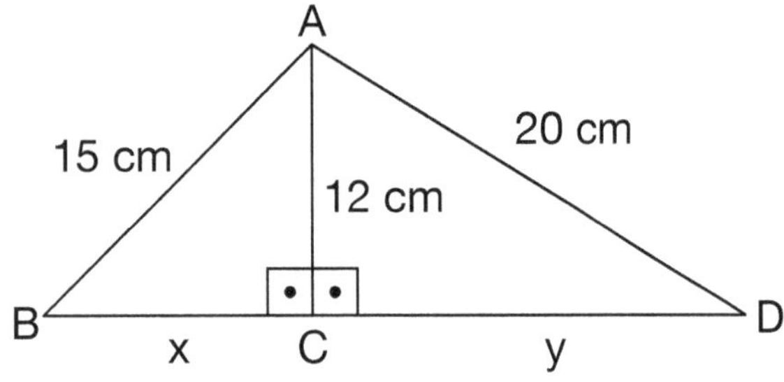

$x^2 + (12 \text{ cm})^2 = (15 \text{ cm})^2$

$x^2 + 144 \text{ cm}^2 = 225 \text{ cm}^2$

$\sqrt{x^2} = \sqrt{81 \text{cm}^2}$

$x = 9 \text{ cm}$

$y^2 + (12 \text{ cm})^2 = (20 \text{ cm})^2$

$y^2 + 144 \text{ cm}^2 = 400 \text{ cm}^2$

$\sqrt{y^2} = \sqrt{256 \text{ cm}^2}$

$y = 16 \text{ cm}$

$x + y = 9 \text{ cm} + 16 \text{ cm}$

$= 25 \text{ cm}$

Correct Answer : D

5. Solution:

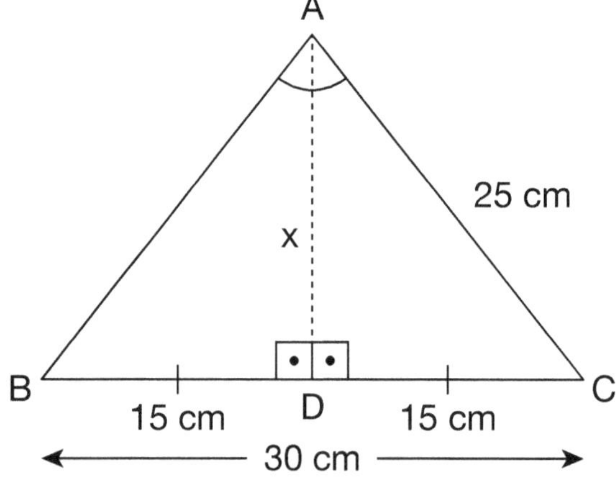

$x^2 + (15 \text{ cm})^2 = (25 \text{ cm})^2$

$x^2 + 225 \text{ cm}^2 = 625 \text{ cm}^2$

$+ \quad -225 \text{ cm}^2 \quad -225 \text{ cm}^2$

$\sqrt{x^2} = \sqrt{400 \text{ cm}^2}$

$x = 20 \text{ cm}$

Correct Answer : D

American Math Academy

MIDPOINT AND DISTANCE TEST SOLUTIONS

1. **Solution:**

Distance formula: $d = \sqrt{(x_2 - x_1)^2 + (y_2 - y_1)^2}$

(9, 7) and (7, 3)

$d = \sqrt{(9-7)^2 + (7-3)^2}$

$d = \sqrt{(2)^2 + (4)^2} = \sqrt{4+16} = \sqrt{20} = 2\sqrt{5}$

Correct Answer : C

2. **Solution:**

Midpoint formula: $M_x = \frac{x_1 + x_2}{2}$, $M_y = \frac{y_2 + y_1}{2}$

(9, 7) and (7, 3)

$M_x = \frac{9+7}{2} = \frac{16}{2} = 8$

$M_y = \frac{7+3}{2} = \frac{10}{2} = 5$

Midpoint is (8, 5)

Correct Answer : B

3. **Solution:**

If the midpoint between A(4, 8) and B(a, b) is (12, 16), then $\frac{4+a}{2} = 12$, 4 + a = 24

a = 20

$\frac{8+b}{2} = 16$, 8 + b = 32, b = 24

Coordinate of B is (20, 24)

Correct Answer : A

American Math Academy

4. **Solution:**

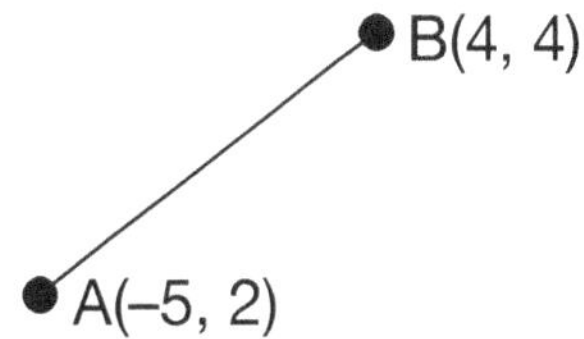

Midpoint of AB = (M_x, M_y)

$M_x = \frac{-5+4}{2} = \frac{-1}{2} = -0.5$

$M_y = \frac{4+2}{2} = \frac{6}{2} = 3$

(–0.5, 3)

Correct Answer : C

5. **Solution:**

Midpoint of (–1, –13) and (9, 23)

$M_x = \frac{-1+9}{2} = \frac{8}{2} = 4$

$M_y = \frac{-13+23}{2} = \frac{10}{2} = 5$

(4, 5)

Correct Answer : A

COORDINATE PLANE TEST SOLUTIONS

1. **Solution:**

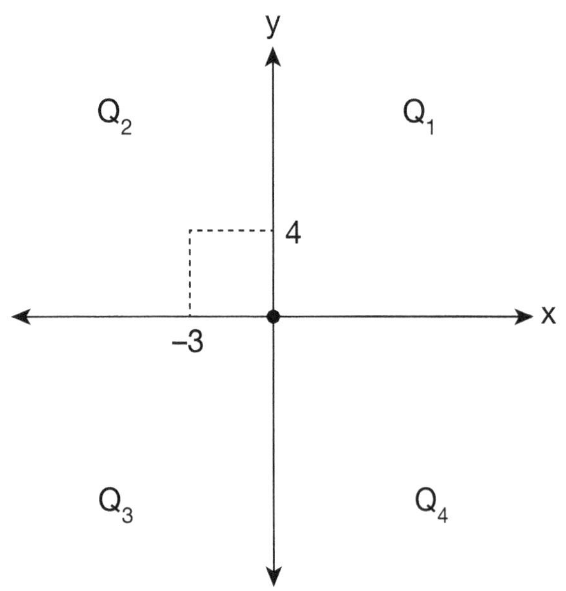

(–3, 4) is located in quadrant II

Correct Answer : B

2. **Solution:**

From all points only choise D is on the y-axis (4, 0)

Correct Answer : D

3. **Solution:**

Distance formula $d = \sqrt{(x_2 - x_1)^2 + (y_2 - y_1)^2}$

(9, 12) and (0, 0)

$d = \sqrt{(9-0)^2 + (12-0)^2} = \sqrt{81+144} = \sqrt{225} = 15$

Correct Answer : D

4. **Solution:**

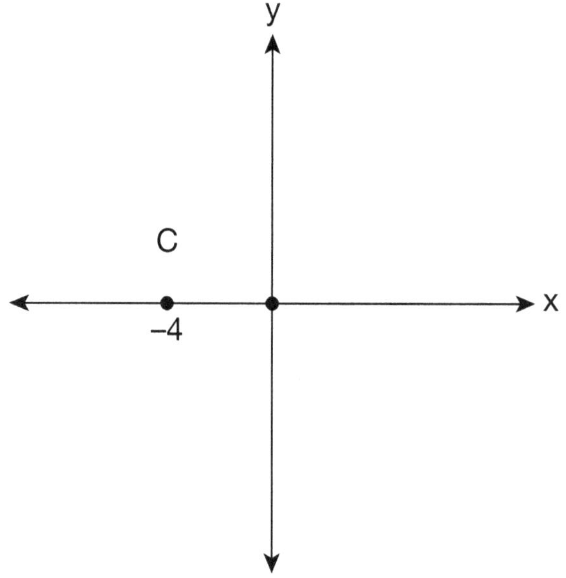

Coordinate of point C is (–4, 0)

Correct Answer : D

5. **Solution:**

No quadrant for point (6, 0), because it's on the x-axis

Correct Answer : D

SLOPE AND SLOPE INTERCEPT FORM TEST SOLUTIONS

1. **Solution:**

A(2, 4) and B(4, 8)

Slope = $m = \frac{Y_2 - Y_1}{X_2 - X_1}$

$m = \frac{8-4}{4-2} = \frac{4}{2} = 2$

If the equation is perpendicular to the line,

then $m_1 = \frac{-1}{m_2}$ $\quad m_2 = \frac{-1}{2}$

from y = mx + b, $y = \frac{-1}{2}x + b$ (Use any point to find b)

$4 = \frac{-1}{2}(2) + b$

$4 = -1 + b$

$5 = b \qquad y = \frac{-1}{2}x + 5$

Correct Answer : B

2. **Solution:**

Coordinate points:

A(1, 3) and B(–1, –3)

slope = $m = \frac{Y_2 - Y_1}{X_2 - X_1} = \frac{-3-3}{-1-1} = \frac{-6}{-2} = 3$

slope = m = 3

Correct Answer : A

3. **Solution:**

Coordinate points: $\left(-2, \frac{1}{3}\right)$ and $\left(1, \frac{-2}{6}\right)$

$m = \frac{Y_2 - Y_1}{X_2 - X_1}$

$m = \frac{\frac{-2}{6} - \frac{1}{3}}{1-(-2)} = \frac{\frac{-1}{3} - \frac{1}{3}}{3} = \frac{-2}{3} \times \frac{1}{3} = \frac{-2}{9}$

Correct Answer : B

4. **Solution:**

Coordinate points: (6 ,3) and (2, k) has a slope of –1

$m = \frac{Y_2 - Y_1}{X_2 - X_1}, -1 = \frac{k-3}{2-6}, -1 = \frac{k-3}{-4},$

then $4 = k - 3 \quad 7 = k$

Correct Answer : C

5. **Solution:**

Coordinate points: (3, 2) and (1, 8)

Slope = $m = \frac{Y_2 - Y_1}{X_2 - X_1} = \frac{8-2}{1-3} = \frac{6}{-2} = -3$

Correct Answer : D

6. **Solution:**

Coordinate points: (0 ,3) and (3, 2)

if y = mx + b, and $m = \frac{Y_2 - Y_1}{X_2 - X_1} = \frac{2-3}{3-0} = \frac{-1}{3}$

$y = \frac{-1}{3}x + b$, since b is y–intercept b = 3

$y = \frac{-1}{3}x + 3$

Correct Answer : C

SIMILARITY THEOREM TEST SOLUTIONS

1. Solution:

From similarity theorem:

$\frac{x}{x+4} = \frac{\cancel{9}^{1}}{\cancel{27}_{3}}$,

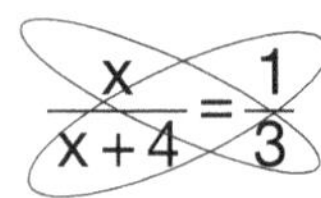

$3x = x + 4$

$2x = 4$

$x = 2$

Correct Answer : A

2. Solution:

From similarity theorem:

$\frac{9}{9+6} = \frac{x}{5}$, $\frac{\cancel{9}^{3}}{\cancel{15}_{5}} = \frac{x}{5}$, $\frac{3}{5} = \frac{x}{5}$

$5x = 15$

$x = 3$

Correct Answer : A

3. Solution:

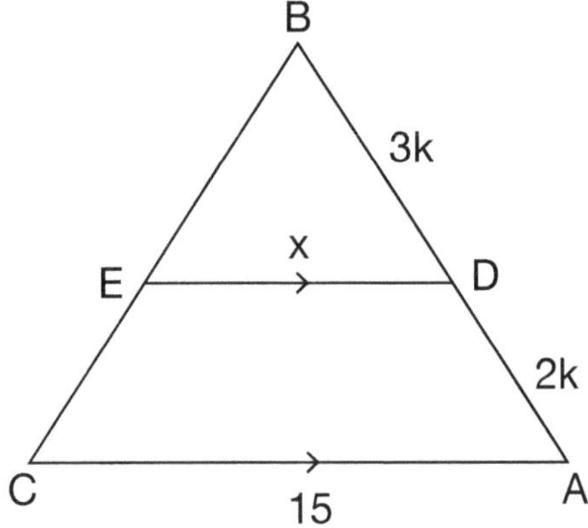

$3|AD| = 2|BD|$ $\quad \frac{x}{15} = \frac{3\cancel{k}}{5\cancel{k}}$

$BD = 3k$ $\quad \frac{x}{15} = \frac{3}{5}$

$AB = 2k$ $\quad 5x = 45, \quad x = 9$

Correct Answer : C

4. Solution:

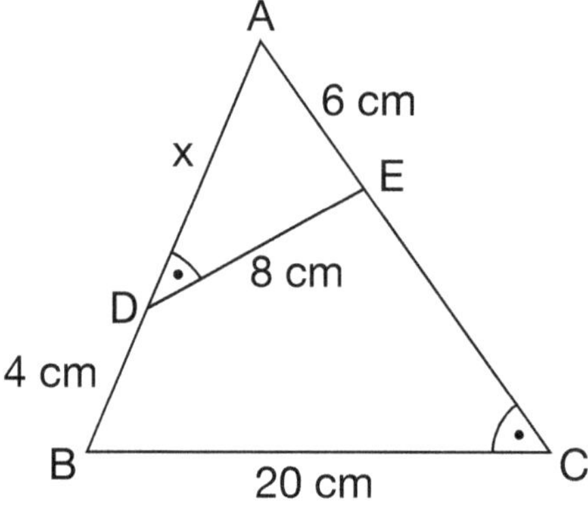

If $\triangle ADE \cong ACB$, then $\frac{AE}{BD} = \frac{DE}{BC}$

$\frac{6}{x+4} = \frac{\cancel{8}^{2}}{\cancel{20}_{5}}$

$\frac{6}{x+4} = \frac{2}{5}$

$2x + 8 = 30$

$2x = 22$

$x = 11$

Correct Answer : C

5. Solution:

From similarity theorem $\frac{DC}{AB} = \frac{KC}{KB}$

$\frac{\cancel{3}^{1}}{\cancel{9}_{3}} = \frac{x}{x+4}$, $\frac{1}{3} = \frac{x}{x+4}$ (Cross multiply)

$3x = x + 4$

$2x = 4$

$x = 2$

Correct Answer : A

6. Solution:

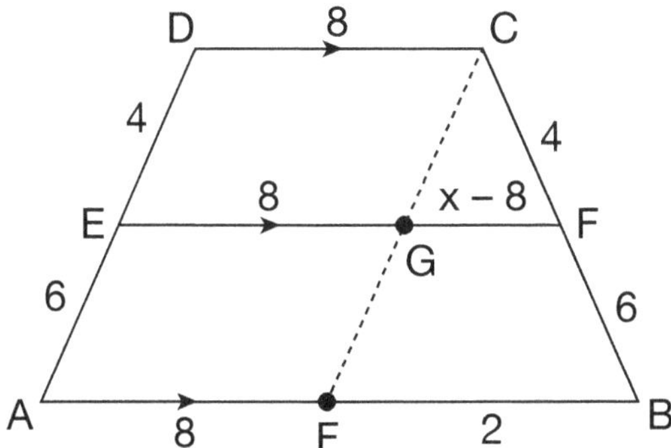

From similerity theorem:

$$\frac{GF}{FB} = \frac{CF}{CB}$$

$$\frac{x-8}{2} = \frac{4}{10}, \ \frac{x-8}{2} = \frac{2}{5}$$

$5(x - 8) = 2 \times 2$

$5x - 40 = 4$

$5x = 44$

$x = 8.8$

Correct Answer : C

AREA AND PERIMETER OF TRIANGLE TEST SOLUTIONS

1. Solution:

Area of $\triangle ABC = \dfrac{3m \times 4m}{2} = \dfrac{12m^2}{2} = 6m^2$

Correct Answer : A

2. Solution:

Area of $\triangle ABC = \dfrac{b \cdot h}{2} = \dfrac{3cm \times 18cm}{2} = 27cm^2$

Correct Answer : B

3. Solution:

Area of equilateral triangle = $\dfrac{a^2\sqrt{3}}{4}$

since each side is 4 Area

$= \dfrac{4^2\sqrt{3}}{4} = \dfrac{16\sqrt{3}}{4} = 4\sqrt{3}$

Correct Answer : C

4. Solution:

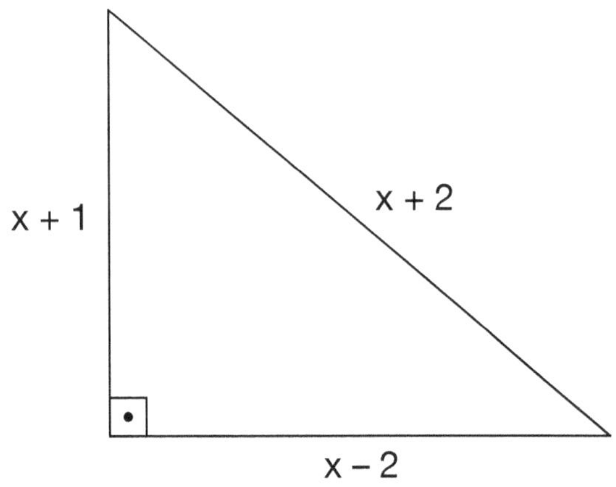

Perimeter = $x + 1 + x + \cancel{2} + x - \cancel{2}$

$P = 3x + 1$

Correct Answer : C

5. Solution:

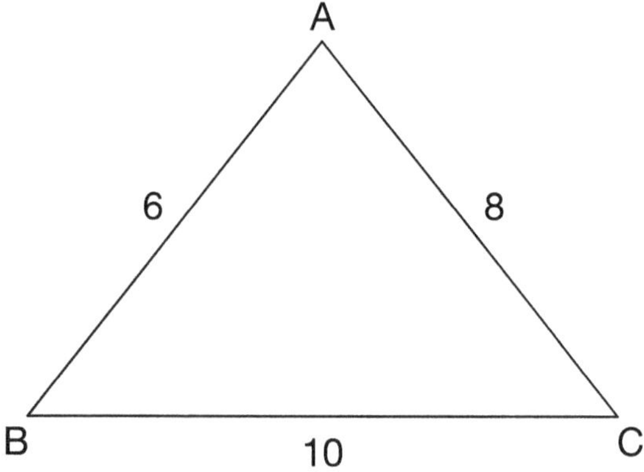

The area of triangle from Heron's Formula:

$a = 10,\ b = 8,\ c = 6$

$s = \dfrac{6+8+10}{2} = \dfrac{24}{2} = 12$

Area = $\sqrt{s(s-a)\ (s-b)\ (s-c)}$

Area = $\sqrt{12(12-10)\cdot(12-8)\ (12-6)}$

Area = $\sqrt{12\cdot 2\cdot 4\cdot 6}$

Area = $\sqrt{24 \times 24}$

Area = $\sqrt{(24)^2} = 24$

Correct Answer : D

AREA AND PERIMETER OF TRIANGLE TEST SOLUTIONS

6. Solution:

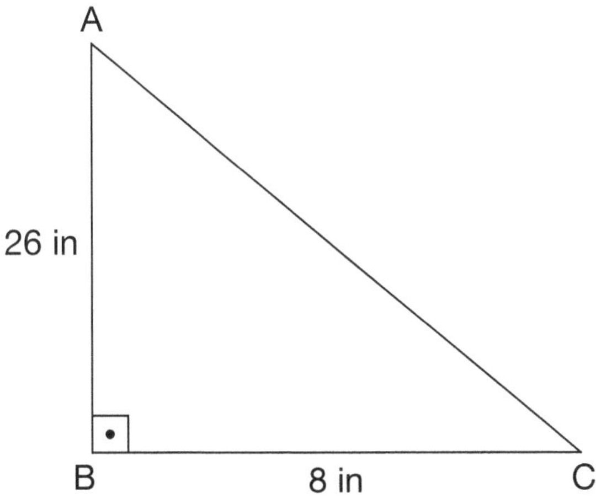

Area of $\triangle$ ABC $= \dfrac{b \times h}{2} = \dfrac{26in \times 8in}{2}$

Area = 104 in^2

Correct Answer : B

7. Solution:

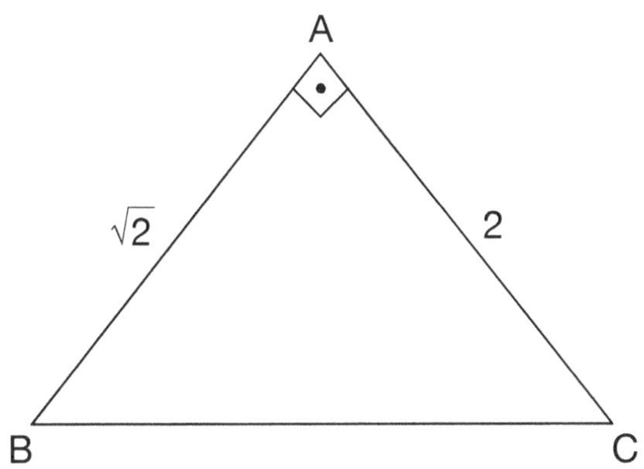

Area of $\triangle$ABC $= \dfrac{\sqrt{2} \times 2}{2} = \dfrac{\cancel{2}\sqrt{2}}{\cancel{2}}$

Area = $\sqrt{2}$

Correct Answer : B

American Math Academy

AREA AND PERIMETER OF QUADRILATERALS TEST SOLUTIONS

1. Solution:

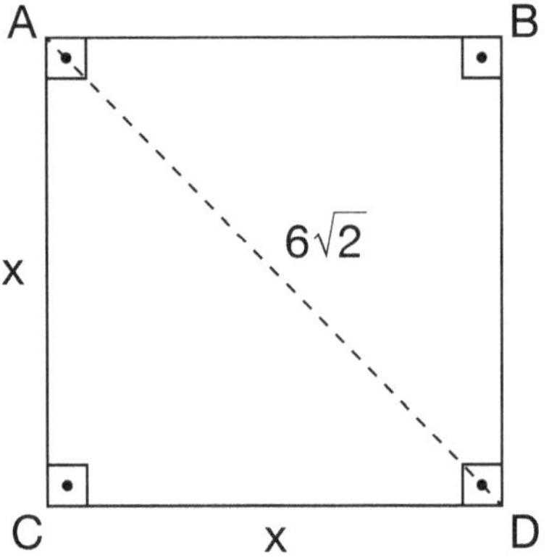

From pythagorean theorem:

$x^2 + x^2 = (6\sqrt{2})^2$

$2x^2 = 72$

$x^2 = 36$

$x = 6$

Area $= x^2 = 36$

Correct Answer : C

2. Solution:

Area of ABCD parallelogram $= b \cdot h$

$= 4 \text{ cm} \times 8 \text{ cm}$

$= 32 \text{ cm}^2$

Correct Answer : A

3. Solution:

If perimeter of rectangle is 48, and $w = 3\ell$ then

$2\ell + 2w = 48$ m, $\ell + w = 24$ m

$\ell + 3\ell = 24$ m

$4\ell = 24$ m

$\ell = 6$ m

$w = 3\ell = 3 \times 6 = 18$ m

Correct Answer : D

American Math Academy

4. Solution:

Area of ABCD trapezoid $= \dfrac{(b_1+b_2)\times h}{2}$

$\text{Area} = \dfrac{(6\,\text{cm}+16\,\text{cm})\times 2\,\text{cm}}{2} = \dfrac{22\,\text{cm}\times \cancel{2}\,\text{cm}}{\cancel{2}}$

Area $= 22 \text{ cm}^2$

Correct Answer : B

5. Solution:

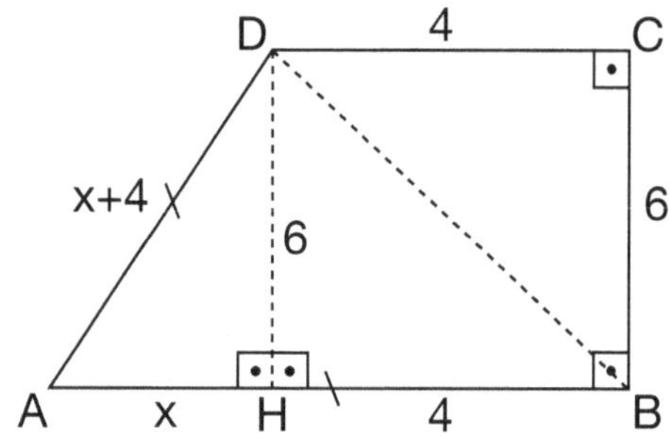

From pythagorean theorem:

$(x+4)^2 = x^2 + 6^2$

$\cancel{x^2} + 8x + 16 = \cancel{x^2} + 36$

$8x + 16 = 36$

$8x = 20 \quad x = \dfrac{5}{2}$

$AB = x + 4 = \dfrac{5}{2} + 4 = \dfrac{13}{2}$

Correct Answer : B

AREA AND PERIMETER OF QUADRILATERALS TEST SOLUTIONS

6. Solution:

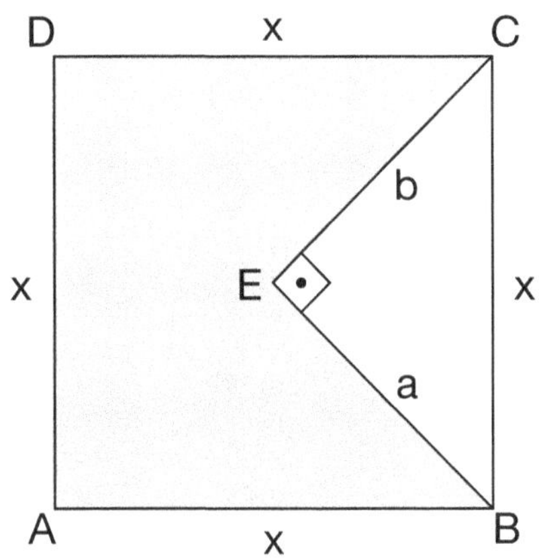

Perimeter of ABCD = 32m

$4x = 32m \quad x = 8m$

Perimeter of $\triangle BEC = 32m = a + b + x = 18m$

From pythagorean theorem:

$a^2 + b^2 = x^2 \quad a^2 + b^2 = 64m^2$

$(a+b)^2 - 2ab = 64$, since $a + b = 10$, then

$100 - 2ab = 64m^2, \quad ab = 18m^2$

$A(BEC) = \dfrac{18m^2}{2} = 9m^2$

Shaded Area $= 64m^2 - 9m^2$

$= 55m^2$

Correct Answer : C

7. Solution:

Area of ABCD trapezoid $= \dfrac{(b_1 + b_2) \times h}{2}$

Area $= \dfrac{(10\,cm + 12\,cm) \times 11cm}{2}$

$= \dfrac{22\,cm \times 11cm}{2} = 121cm^2$

Correct Answer : D

8. Solution:

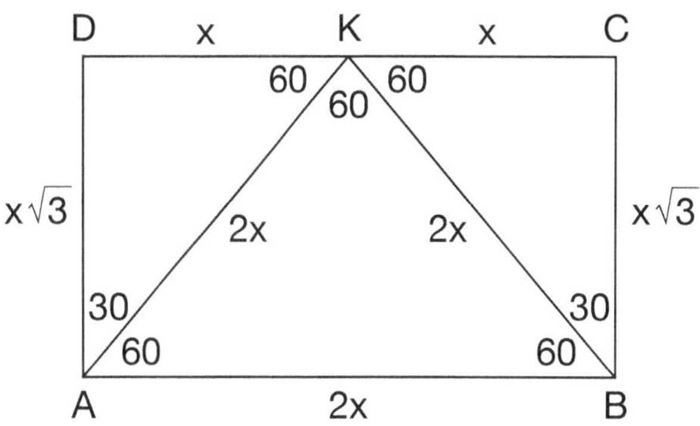

Since the area of rectangle ABCD is $8\sqrt{3}$ and $\triangle AKB$ is an equilateral triangle,

$A(ABCD) = 8\sqrt{3}$

$x\sqrt{3} \cdot 2x = 8\sqrt{3}$

$2x^2 = 8$

$x^2 = 4$

$x = 2$, $AB = 2x$

$AB = 2 \cdot 2 = 4$

Correct Answer : B

TRANSFORMATIONS TEST SOLUTIONS

1. **Solution:**

   From coordinate grid point M is (3, 3)

   after point M reflected over the x–axis

   $M' = (3,-3)$

   Correct Answer : D

2. **Solution:**

   A(–2,–4) reflected over the y-axis is

   $A' = (2,-4)$

   Correct Answer : B

3. **Solution:**

   From (–4,5) to (6,–3)

   x coordinate 10 units right

   y coordinate 8 units down

   Correct Answer : D

4. **Solution:**

   From coordinate grid point A is (2,1)

   If A gets translated 5 units down,

   then $A' = (2,-4)$

   Correct Answer : C

5. **Solution:**

   A(3,5) reflects across to the origin

   $A' = (-3,-5)$

   Correct Answer : C

CIRCLES TEST SOLUTIONS

1. **Solution:**

   r = 13ft $\quad C = 2\pi r$

   $C = 2\pi \times 13\text{ft} = 26\pi \text{ ft}$

   Correct Answer : B

2. **Solution:**

   $\text{Area} = \pi r^2$ $\quad$ If d = 16ft, then r = 8ft

   $\text{Area} = \pi(8\text{ft})^2$

   $= 64\pi \text{ ft}^2$

   Correct Answer : D

3. **Solution:**

   Area of circle A $= \pi r^2 = \pi(4\text{ft})^2 = 16\pi\text{ft}^2$

   Area of circle B $= \pi r^2 = \pi(10\text{ft})^2 = 100\pi\text{ft}^2$

   $$\frac{\text{Area of circle A}}{\text{Area of circle B}} = \frac{16\pi \text{ ft}^2}{100\pi \text{ ft}^2} = \frac{16}{100}$$

   $$= \frac{4}{25}$$

   Correct Answer : A

4. **Solution:**

   Area of shaded circle

   $$\frac{a \times \pi \times r^2}{360^\circ} \qquad \alpha^\circ = 30^\circ$$

   $r = 4$ cm

   Area of shaded part

   $$= \frac{\overset{1}{\cancel{30^\circ}} \times \pi \times 16\,\text{cm}^2}{\underset{12}{\cancel{360^\circ}}} = \frac{16\pi}{12}\text{cm}^2$$

   $$= \frac{4}{3}\pi\,\text{cm}^2$$

   Correct Answer : B

5. **Solution:**

   $$\frac{\text{Circumference}}{\text{Area}} = \frac{2\cancel{\pi r}}{\cancel{\pi} r^{\cancel{2}}}, \text{ then } \frac{2}{r} = \frac{4}{6}$$

   $4r = 12$

   $r = 3$

   Correct Answer : B

6. Solution:

$$\frac{\text{Circumference of circle X}}{\text{Circumference of circle Y}}$$

$$= \frac{2\pi \times 3\,ft}{2\pi \times 8\,ft} = \frac{3}{8}$$

Correct Answer : A

7. Solution:

$6x + 3x + 5x + 4x = 360°$

$18x = 360°$

$x = 20°,\ a = \frac{3x}{2} = \frac{3x\,20°}{2} = 30°$

Correct Answer : C

8. Solution:

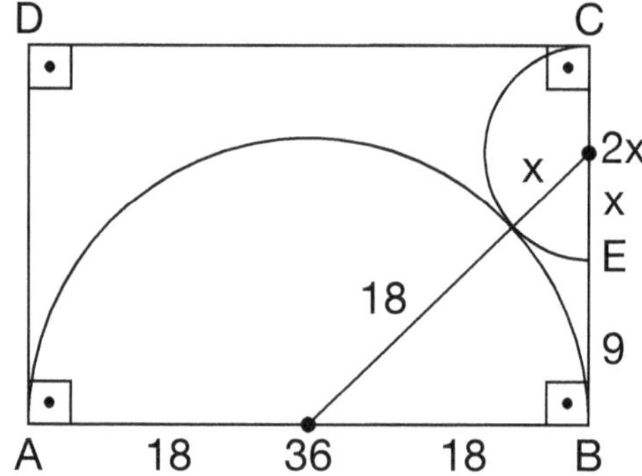

From pythagorean theorem:

$(x + 9)^2 + 18^2 = (x + 18)^2$

$\cancel{x^2} + 18x + 81 + \cancel{324} = \cancel{x^2} + 36x + \cancel{324}$

$81 = 18x$

$\frac{81}{18} = x$

$\frac{9}{2} = x \quad , \quad 4.5 = x$

Correct Answer : C

TRIGONOMETRY TEST SOLUTIONS

1. Solution:

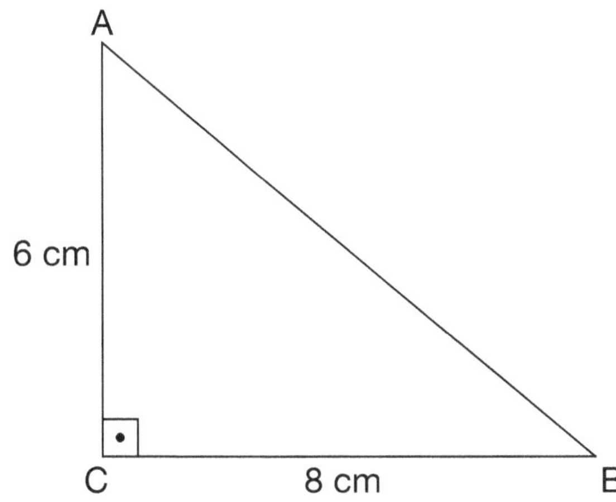

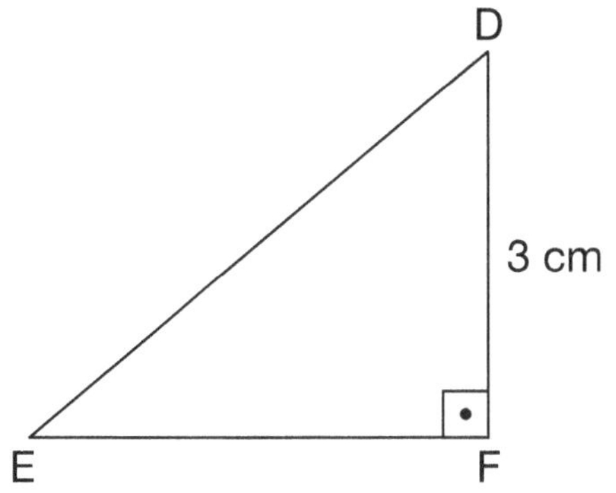

If tanA = CotD

$\tan A = \frac{8}{6}$, $\cot D = \frac{3}{EF}$ $\Big\}$ $\frac{8}{6} = \frac{3}{EF}, EF = \frac{18}{8} = \frac{9}{4}$

Correct Answer : D

2. Solution:

$\frac{60°}{180°} = \frac{R}{\pi}$, $R = \frac{1}{3}\pi$

Correct Answer : A

3. Solution:

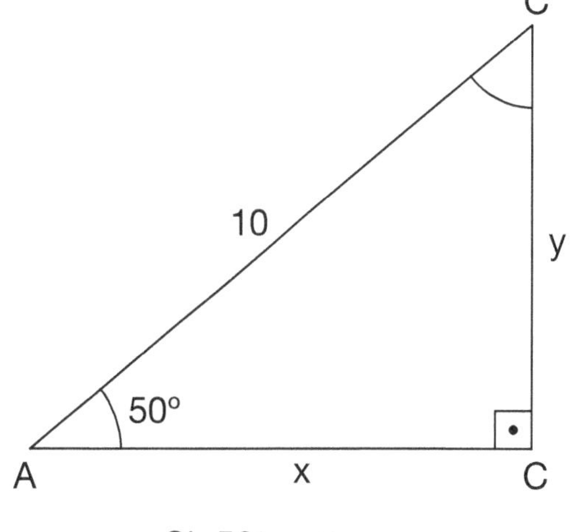

$\tan 50° = \frac{\sin 50°}{\cos 50°} = \frac{y}{x}$

Correct Answer : B

4. Solution:

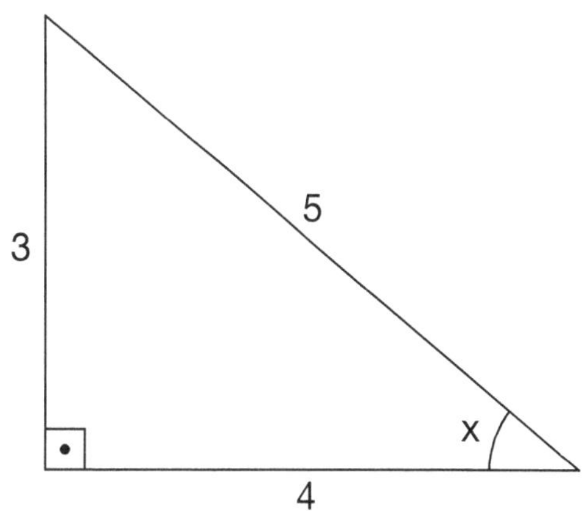

$\cot x = \frac{\text{Adj}}{\text{Opposite}}$

$\cot x = \frac{4}{3}$

Correct Answer : B

American Math Academy

5. Solution:

$\cos30° \times \sin60° \times \tan60°$

$= \frac{\sqrt{3}}{2} \times \frac{\sqrt{3}}{2} \times \sqrt{3}$

$= \frac{3\sqrt{3}}{4}$

Correct Answer : D

6. Solution:

$\frac{\sin x}{\cos x} + \frac{\cos x}{1+\sin x} = \frac{\sin x + \sin^2 x + \cos^2 x}{\cos x(1+\sin x)}$

$= \frac{\cancel{1+\sin x}}{\cos x(\cancel{1+\sin x})} = \frac{1}{\cos x} = \sec x$

Correct Answer : A

7. Solution:

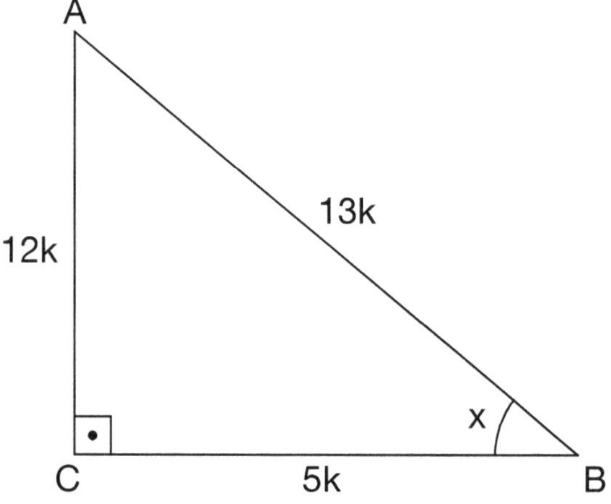

$\tan(90 - x°) = \cot x° = \frac{5k}{12k} = \frac{5}{12}$

Correct Answer : B

American Math Academy

VOLUME TEST SOLUTIONS

1. Solution:

$V_{cone} = \frac{1}{3}\pi r^2 h$ $\quad$ $V_{cone} = V_{sphere}$

$V_s = \frac{4}{3}\pi r^3$ $\quad$ $\frac{1}{3}\pi r^2 h = \frac{4}{3}\pi r^3$

$\frac{1}{4}h = r$

Correct Answer : D

2. Solution:

Suppose $r = 4$ $\quad$ $V = \pi r^2$

$h = 5$ $\quad$ $V = \pi \cdot 16 \cdot 5 = 80\pi$

$r \rightarrow 25\%$ increase $r = 5$

$r \rightarrow 20\%$ decrease $h = 4$

$v = \pi r^2 = 25 \times 4 \times \pi = 100\pi$

$\frac{100\pi - 80\pi}{80\pi} = \frac{1}{4} = 25\%$ increase

Correct Answer : C

3. Solution:

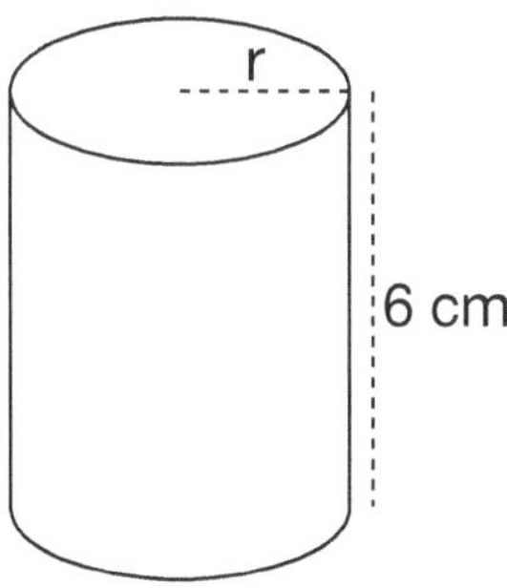

Volume of cylinder $= \pi r^2 h$

$\pi r^2 \times 6\text{ cm} = 72\pi\text{ cm}^3$

$r^2 \times 6\text{ cm} = 72\text{ cm}^3$

$r^2 = \frac{72\text{ cm}^3}{6\text{ cm}}$, $r^2 = 12\text{ cm}^2$

$r = 2\sqrt{3}\text{ cm}$

Correct Answer : B

4. Solution:

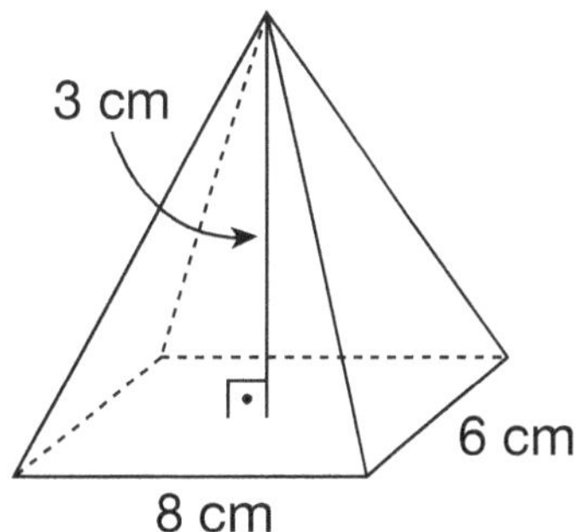

$V_{pyramid} = \frac{l \times w \times h}{3}$

$V = \frac{3\text{ cm} \times 6\text{ cm} \times 8\text{ cm}}{3}$

$V = 48\text{ cm}^3$

Correct Answer : C

VOLUME TEST SOLUTIONS

5. Solution:

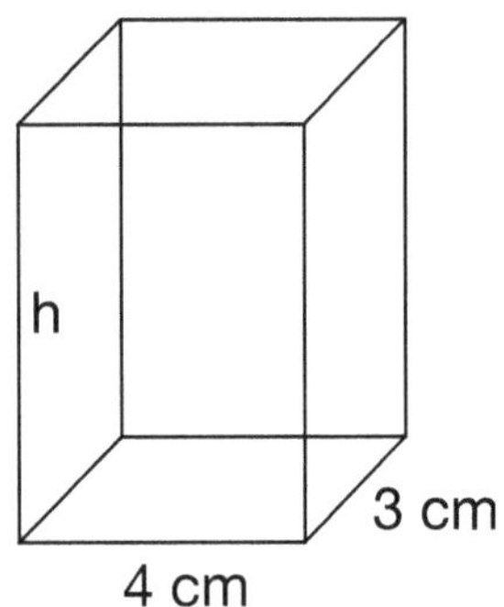

$V_{rectangular} = l \times w \times h$

$h \times 4\text{ cm} \times 3\text{ cm} = 96\text{ cm}^3$

$h = \dfrac{96\text{ cm}^3}{12\text{ cm}^3}$

$h = 8\text{ cm}$

Correct Answer : B

6. Solution:

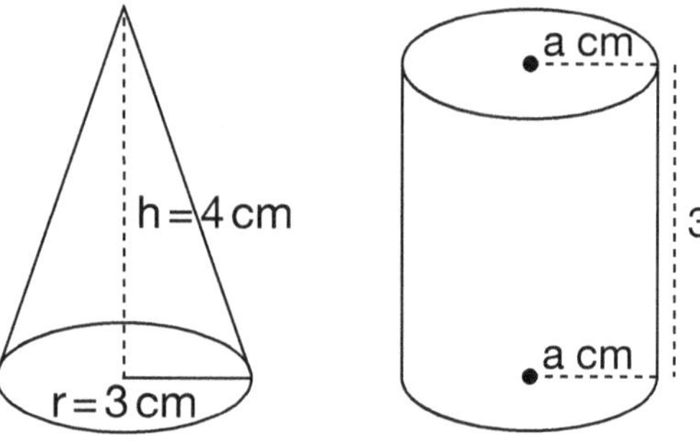

$V_{cone} = \dfrac{1}{3}\pi r^2 h = \dfrac{1}{3}\pi \times 9 \times 4 = 12\pi$

$V_{cylinder} = \pi r^2 h = \pi \times a^2 \times 3$

$= \pi \times a^2 \times 3$

$V_{cone} = V_{cylinder}$

$12\pi = \pi \times a^2 \times 3$

$\dfrac{12}{3} = a^2$, $a^2 = 4$, $a = 2\text{ cm}$

Correct Answer : B

MIXED REVIEW TEST 1 SOLUTIONS

1. Let the common ratio be x.

   Therefore, $4x + 5x = 90°$, $x = 10°$.

   Smaller angle = $4x = 40°$.

   Correct answer: B

2. Exterior angle + Interior angle = 180°.

   Therefore, Exterior angle = 180° - 40° = 140°.

   Correct answer: B

3. Scale factor = New length / Original length = $\frac{10}{4} = 2.5$.

   Correct answer: A

4. If two triangles are congruent by SAS, ASA can be deduced.

   Correct answer: B

5. Diagonal length = $\sqrt{(\text{length}^2 + \text{width}^2)}$

   $= \sqrt{(12^2 + 9^2)} = 15$ cm.

   Correct answer: B

6. Sum of interior angles

   $= (n - 2) \times 180° = (12 - 2) \times 180° = 1800°$.

   Correct answer: A

7. Ratio of areas of similar triangles = (ratio of corresponding sides)^2 = $(3:5)^2 = 9:25$.

   Correct answer: B

8. The other acute angle

   $= 90° - 30° = 60°$. $\text{Sin}(60°) = \frac{\sqrt{3}}{2}$.

   Correct answer: C

9. $\cos(\theta) = \sqrt{(1 - \sin^2(\theta))}$

   $= \sqrt{\left(1 - \left(\frac{5}{13}\right)^2\right)} = \frac{12}{13}$.

   Correct answer: A

10. Area = $\frac{1}{2} \times$ (sum of bases) $\times$ height

    $= \frac{1}{2} \times (10 \text{ cm} + 15 \text{ cm}) \times 8 \text{ cm} = 100 \text{ cm}^2$.

    Correct answer: A

MIXED REVIEW TEST 1 SOLUTIONS

11. Total Surface Area = $2\pi r(h + r) = 2\pi(3 \text{ cm})(10 \text{ cm} + 3 \text{ cm}) = 2(3.14)(3)(13)$

$= 244.92 \text{ cm}^2$

Correct answer: A

12. Volume = $\left(\frac{1}{3}\right)\pi r^2 h = \left(\frac{1}{3}\right)\pi(4 \text{ cm})^2(9 \text{ cm})$

$= 48\pi \text{ cm}^3$.

Correct answer: A

13. Circumference = $2\pi r$ = 31.4 cm

$2(3.14)(r) = 31.4$

$r = 5$ cm

Correct answer: A

14. The sum of exterior angles of a polygon is always 360°.

Correct answer: D

15. The slope of a line perpendicular to L_2 is the negative reciprocal of the slope of L_1, hence $-\frac{1}{3}$.

Correct answer: C

16. When a point is reflected over the x-axis, only the y-coordinate changes sign.

Correct answer: B

17. The Hypotenuse-Leg (HL) Theorem states that right triangles are congruent if the hypotenuse and one leg of one triangle are congruent to the hypotenuse and one leg of another triangle.

Correct answer: D

18. Adjacent angles in a parallelogram are supplementary, therefore the measure of the adjacent angle is 180° - 60° = 120°.

Correct answer: C

19. The formula for the measure of each interior angle of a regular polygon is $(n - 2)\frac{180°}{n}$, where n is the number of sides. For a hexagon, n = 6.

So, interior angle = $\frac{(4 \cdot 180°)}{6} = 120°$.

Correct answer: A

MIXED REVIEW TEST 1 SOLUTIONS

20. The perimeter of the larger triangle is $\left(\frac{3}{2}\right)$ times the perimeter of the smaller triangle, 50, $30\text{cm}\left(\frac{3}{2}\right) = 45\text{cm}$.

Correct answer: A

21. By the Pythagorean theorem,

hypotenuse = $\sqrt{(7^2 + 24^2)}$ = 25 cm.

Correct answer: A

22. Tan(θ) = Opposite/Adjacent = $\frac{3}{4}$,

$\text{Cosine}\theta = \frac{\text{adj}}{\text{hypotenuse}} = \frac{4}{5}$

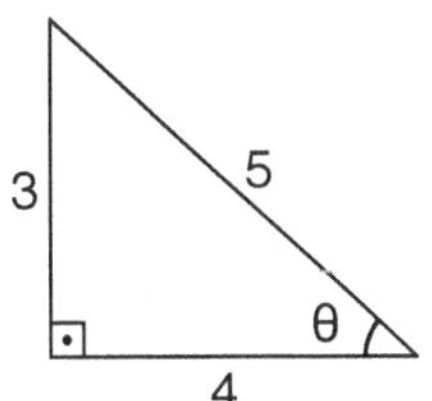

Correct answer: A

23. Area of sector =

$$\left(\frac{\text{central angle}}{360}\right)\cdot\pi r^2 = \left(\frac{90°}{360°}\right)\cdot\pi(10\,\text{cm})^2$$
$$= \frac{100\pi\text{cm}^2}{4}$$
$$= 25\pi\text{cm}^2$$

Correct answer: A

24. Total Surface Area

= 2(lw + lh + wh)

= 2(4·5 + 4·6 + 5·6)

= 2(20 + 24 + 30)

= 2(74)

= 148 cm².

Correct answer: D

MIXED REVIEW TEST 2 SOLUTIONS

1. A line segment is defined by two endpoints.

Correct answer: B

2. The point (0,0) is where the x-axis and y-axis intersect, called the origin.

Correct answer: A

3. A line that intersects two or more lines at distinct points is called a transversal.

Correct answer: B

4. The slope of a line perpendicular to another line is the negative reciprocal of the original line's slope. $m = \frac{-1}{4}$

Correct answer: D

5. Reflection, rotation, and translation preserve the size and shape of the object, but dilation changes the size.

Correct answer: D

6. Dilation is the transformation that enlarges or reduces a figure.

Correct answer: C

7. This is the Side-Angle-Side (SAS) Postulate of congruence.

Correct answer: A

8. In congruent triangles, corresponding angles are equal.

Correct answer: C

9. The sum of interior angles of a quadrilateral is 360°.

Correct answer: C

10. A quadrilateral with exactly one pair of parallel sides is a trapezoid.

Correct answer: C

11. The sum of interior angles of a hexagon is $(6 - 2) \cdot 180° = 720°$.

Correct answer: B

12. In a regular polygon, all sides and all angles are equal.

Correct answer: A

13. Two triangles are similar if their corresponding sides are proportional and corresponding angles are equal.

Correct answer: C

MIXED REVIEW TEST 2 SOLUTIONS

14. The perimeter of the larger triangle is $\left(\frac{3}{2}\right)$ times the perimeter of the smaller triangle, so 30 cm$\cdot\left(\frac{3}{2}\right)$ = 45 cm.

Correct answer: A

15. In a right triangle, the side opposite the right angle is the hypotenuse.

Correct answer: A

16. The other acute angle in a right triangle is $90° - 30° = 60°$.

Correct answer: C

17. In a right triangle, sin(θ) is the length of the side opposite θ divided by the length of the hypotenuse.

Correct answer: C

18. The cosine of a 60° angle is $\frac{1}{2}$.

Correct answer: A

19. The formula for the area of a trapezoid is $\frac{1}{2}\cdot$(sum of bases)$\cdot$height.

Correct answer: B

20. Area of a rectangle is given by length × width. So, 48 cm^2 = 8 cm × width, which implies width = 6 cm.

Correct answer: B

21. The formula for the surface area of a cylinder is $2\pi r(h + r)$.

Correct answer: A

22. The total surface area of a rectangular prism is

$2(lw + lh + wh) = 2(3\cdot4 + 3\cdot5 + 4\cdot5)$

$= 2(12 + 15 + 20)$

$= 2(47)$

$= 94$ cm^2.

Correct answer: A

23. The formula for the volume of a sphere is $\frac{4}{3}\pi r^3$.

Correct answer: A

24. The formula for the volume of a cylinder is $\pi r^2h = \pi(3\text{ cm})^2(4\text{ cm}) = 36\pi\text{ cm}^3$.

Correct answer: A

MIXED REVIEW TEST 3 SOLUTIONS

1. The ratio of the areas of two similar triangles is the square of the ratio of their corresponding side lengths. Since the ratio of AB to DE is 3 : 5, the ratio of the areas will be $\left(\frac{3}{5}\right)^2 = 9:25$.

   Correct answer: C

2. According to the Alternate Interior Angles Theorem, the lines are parallel.

   Correct answer: B

3. A 90° counterclockwise rotation about the origin takes the point (x, y) to (-y, x).

   Correct answer: B

4. According to the definition of congruent triangles, their corresponding parts are congruent.

   Correct answer: A

5. In a rectangle, the diagonals are congruent but not perpendicular.

   Correct answer: B

6. The sum of the exterior angles of a polygon, one at each vertex, is always 360°.

   Correct answer: C

7. Two triangles are similar if they have three pairs of proportional sides.

   Correct answer: C

8. In a 45 - 45 - 90 right triangle, the ratio of the length of the hypotenuse to the length of a leg is $\sqrt{2}:1$.

   Correct answer: C

9. Using the Pythagorean identity

   $\sin^2(\theta) + \cos^2(\theta) = 1$,

   we find that $\cos(\theta)$

   $= \sqrt{(1 - \sin^2(\theta))} = \sqrt{\left(1 - \left(\frac{3}{5}\right)^2\right)} = \frac{4}{5}$.

   Correct answer: A

10. The area of a rhombus is given by the formula $\left(\frac{1}{2}\right)d_1 \cdot d_2 = \left(\frac{1}{2}\right)10 \cdot 24 = 120\text{ cm}^2$.

    Correct answer: A

11. The total surface area of a sphere is given by the formula $4\pi r^2 = 4\pi(5\text{ cm})^2 = 100\pi\text{ cm}^2$.

    Correct answer: C

12. The volume of a cone is given by the formula $\left(\frac{1}{3}\right)\pi r^2 h = \left(\frac{1}{3}\right)\pi(3\text{ cm})^2(4\text{ cm}) = 12\pi\text{ cm}^3$.

    Correct answer: A

13. The equation of a circle with center (h, k) and radius r is given by $(x - h)^2 + (y - k)^2 = r^2$. Plugging in the values from the question, we get $(x - 2)^2 + (y + 3)^2 = 5^2$.

    Correct answer: A

MIXED REVIEW TEST 3 SOLUTIONS

14. The measure of an angle formed by two tangent lines to a circle that intersect outside the circle is half the difference of the intercepted arcs.

Correct answer: A

15. The slope of the line $3x - 4y = 7$ is $\frac{3}{4}$, so the slope of a line perpendicular to it is $-\frac{4}{3}$.

Correct answer: C

16. The matrix $\begin{pmatrix} 0 & -1 \\ 1 & 0 \end{pmatrix}$ represents a 90° rotation counterclockwise about the origin.

Correct answer: C

17. In congruent triangles, corresponding medians are congruent.

Correct answer: C

18. In a parallelogram, opposite angles are equal.

Correct answer: C

19. The interior angle of a regular pentagon is given by the formula $\frac{(n-2)\cdot 180}{n}$

$\frac{(5-2)\cdot 180}{5} = 108°$ where n is the number of sides. Plugging n = 5, we get 108°.

Correct answer: A

20. Two triangles are similar if their corresponding sides are in proportion and their corresponding angles are equal.

Correct answer: A

21. The Pythagorean Theorem is applicable only to right-angled triangles.

Correct answer: B

22. In a right-angled triangle, the tangent of angle θ is the ratio of the opposite side to the adjacent side.

Correct answer: A

23. The area of an equilateral triangle with side length a is given by the formula $\frac{(a^2\sqrt{3})}{4}$.

Correct answer: A

24. The total surface area of a cylinder is given by the formula $2\pi r(h + r)$.

Correct answer: A

MIXED REVIEW TEST 4 SOLUTIONS

1. Since the ratio of AB to DE is 5:6, we can set up the proportion $\frac{5}{6} = \frac{9}{DE}$. Solving for DE, we get $DE = \left(\frac{6}{5}\right).9 = 10.8\,\text{cm}$.

Correct answer: A

2. A circle has 360°. An arc that measures 90° is $\frac{1}{4}$th of a circle.

Correct answer: A

3. Arc length

$= \left(\frac{\theta}{360}\right)\cdot(2\pi r)$

$= \left(\frac{60°}{360}\right)\cdot(2\pi(10\text{ cm}))$

$= \left(\frac{1}{6}\right)\cdot 20\pi\text{ cm} = \frac{10\pi}{3}\text{ cm}.$

Correct answer: D

4. A full circle is equal to 2π radians.

Correct answer: B

5. Arc length $= \theta\cdot r = \left(\frac{\pi}{3}\right)\cdot(6\text{ cm}) = 2\pi\text{ cm}.$

Correct answer: B

6. Area of sector $= \left(\frac{\theta}{360}\right)\cdot\pi r^2$

$= \left(\frac{45°}{360}\right)\cdot\pi(8\text{ cm})^2$

$= \pi\frac{64\text{cm}^2}{8}$

$= 8\pi\text{ cm}^2.$

Correct answer: A

7. An inscribed angle that subtends a diameter of a circle is 90°.

Correct answer: C

8. The circumradius of an equilateral triangle with side length "a" is given by the formula $r = \frac{a}{\sqrt{3}}$.

Plugging a = 6 cm gives :

$r = \frac{6\text{ cm}}{\sqrt{3}} = 2\sqrt{3}\text{ cm}.$

Correct answer: C

9. Two tangents drawn from an external point to a circle are congruent.

Correct answer: B

MIXED REVIEW TEST 4 SOLUTIONS

10. Given,

Radius of circle = 12 cm

For Equilateral triangle,

Radius of circumcircle

$= \frac{\text{side of Equilateral triangle}}{\sqrt{3}}$

Side of an Equilateral triangle = $\frac{12}{\sqrt{3}}$ cm

Area of Equilateral triangle

$= \left(\frac{\sqrt{3}}{4}\right) \cdot \text{side.side}$

$= \left(\frac{\sqrt{3}}{4}\right) \cdot \frac{12}{\sqrt{3}} \cdot \frac{12}{\sqrt{3}}$

$= \frac{108}{\sqrt{3}}$ sq.cm

Correct answer: D

11. The standard equation of a circle with center (h, k) and radius r is given by

$(x - h)^2 + (y - k)^2 = r^2$.

Plugging h = 2, k = 3, and

r = 5 gives $(x - 2)^2 + (y - 3)^2 = 25$.

Correct answer: A

12. Completing the square on the given equation: $x^2 + y^2 + 4x - 6y - 12 = 0$ gives $(x + 2)^2 + (y - 3)^2 = 25$. Hence, the center of the circle is (–2, 3).

Correct answer: A

13. The slope-intercept form of a linear equation is y = mx + b, where m represents the slope and b represents the y-intercept. y = 2x + 3

Correct answer: A

14. The statement $p \wedge -p$ is a contradiction, as it's saying p is both true and false at the same time.

Correct answer: B

15. The contrapositive of $p \longrightarrow q$ is $-q \longrightarrow -p$.

Correct answer: C

16. The biconditional $p \longleftrightarrow q$ is true only when p and q have the same truth values.

Correct answer: A

17. Rigid transformations such as translations, rotations, and reflections preserve both distance and angle measures.

All of the above.

Correct answer: D

18. To translate the point (3, 4)5 units to the right, you add 5 to the x-coordinate. So, the x-coordinate becomes:

$3 + 5 = 8$

To translate the point (3, 4)3 units up, you add 3 to the y-coordinate. So, the y-coordinate becomes :

$4 + 3 = 7$

Therefore, the translated point is (8, 7)

Correct answer: A

19. When rotating a point 90° counterclockwise about the origin, the new coordinates become (–y, x).

Correct answer: A

20. Reflecting a point over the x-axis negates the y-coordinate while keeping the x-coordinate the same.

Correct answer: A

21. Dilation does not preserve distance; it enlarges or reduces a figure proportionally, making it a non-rigid transformation.

Correct answer: D

22. Dilation with a scale factor of 2 doubles the coordinates of the original point.

Correct answer: A

23. Rotational symmetry is about an object looking the same after a certain amount of rotation less than 360°.

Correct answer: B

FINAL TEST SOLUTIONS

1. Solution:

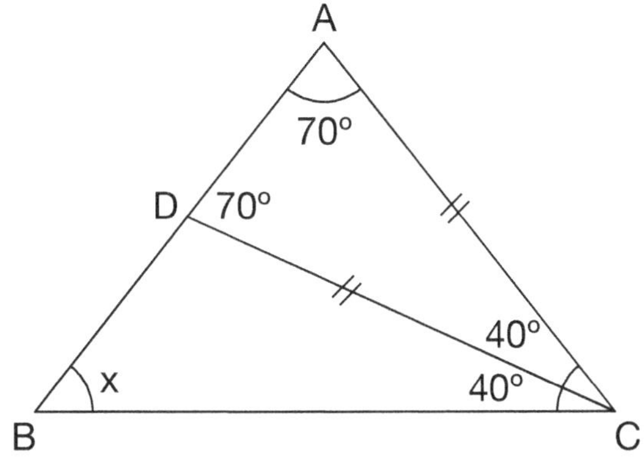

$x + 40^\circ = 70^\circ$

$x = 30^\circ$

Correct Answer : A

2. Solution:

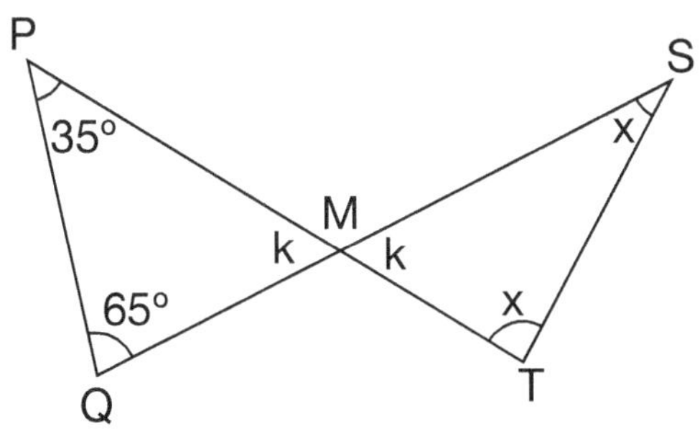

$k + 65^\circ + 35^\circ = 180^\circ$

$k + 100^\circ = 180^\circ$

$k = 180^\circ$

$2x + k = 180^\circ$

$2x + 80^\circ = 180^\circ$

$2x = 10^\circ$

$x = 50^\circ$

Correct Answer : D

3. Solution:

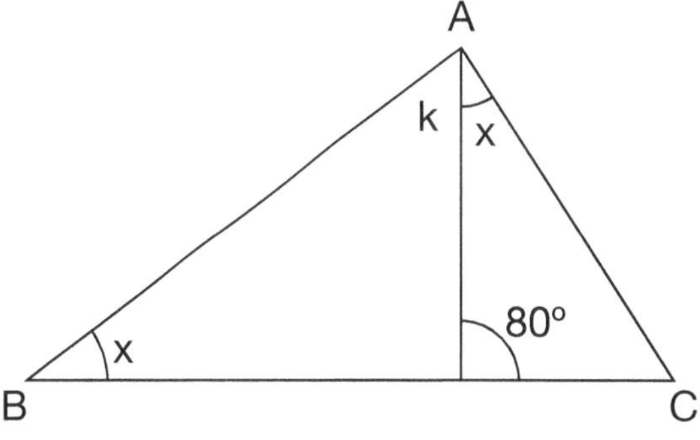

$80^\circ = x + k$

Angle of $\triangle(BAC) = x + k = 80^\circ$

Correct Answer : C

4. Solution:

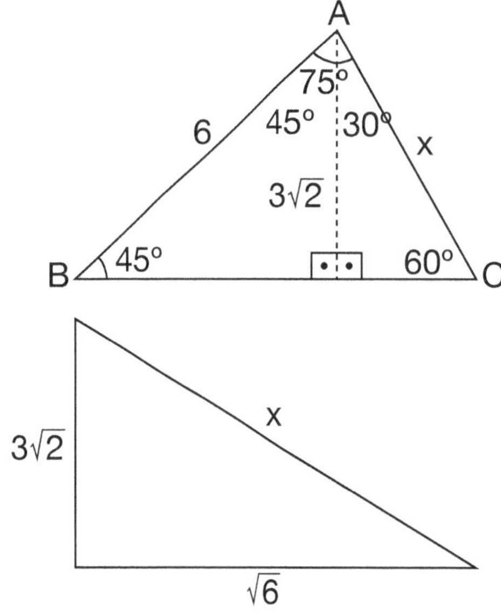

$x^2 = (3\sqrt{2})^2 + (\sqrt{6})^2$

$x^2 = 18 + 6$

$x^2 = 24$

$x = 2\sqrt{6}$

Correct Answer : C

5. Solution:

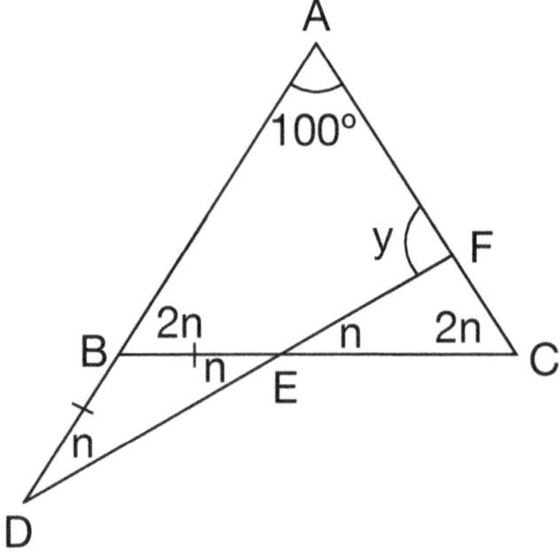

$2n + 2n + 100° = 180°$

$4n + 100° = 180°$

$4n = 80°$

$n = 20°$

$y = 3n$

$y = 60°$

Correct Answer : D

6. Solution:

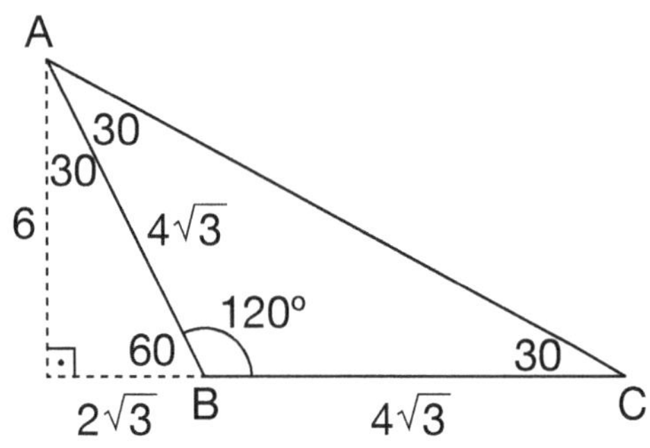

$30° \longrightarrow 2\sqrt{3}$

$60° \longrightarrow 6$

$\text{Area} = \frac{h \cdot b}{2} = \frac{6 \cdot 4\sqrt{3}}{2}$

$= 12\sqrt{3}$

Correct Answer : A

7. Solution:

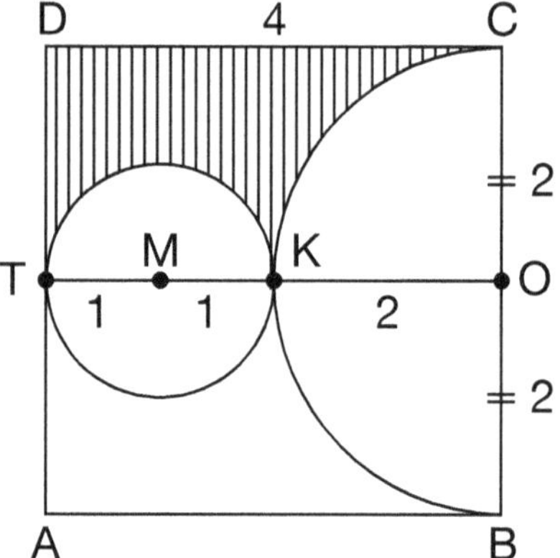

$A(ABCD) = 4^2 = 16ft^2$

$A(TDCO) = 4 \cdot 2 = 8ft^2$

Area of small half circle $= \frac{\pi r^2}{2} = \frac{\pi}{2}$

Area of quarter of big circle $= \frac{\pi r^2}{2} = \frac{\pi \cdot 2^2}{4} = \pi$

Shaded area $= 8 - \left(\frac{\pi}{2} + \pi\right) = 8 - \frac{3\pi}{2}$

Correct Answer : A

FINAL TEST SOLUTIONS

8. Solution:

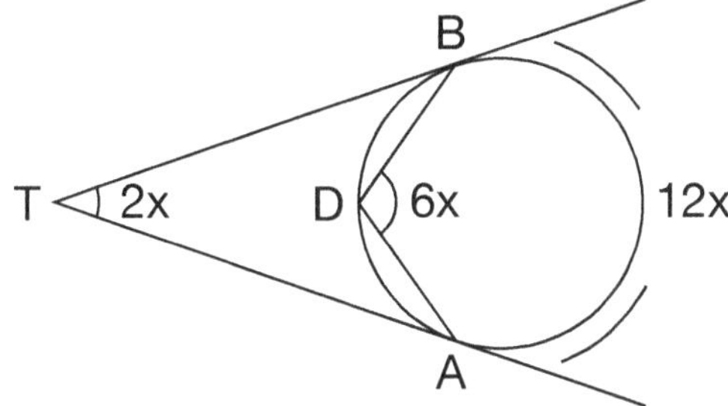

ArcADB = 180° – 2x

ArcAB = 12x

12x + 180° – 2x = 360°

10x – 180° = 360°

10x = 180°

x = 18°

Correct Answer : C

9. Solution:

4k + 5k + 9k = 180°

18k = 180°

k = 10°

smallest angle = 4k

= 40°

Correct Answer : A

10. Solution:

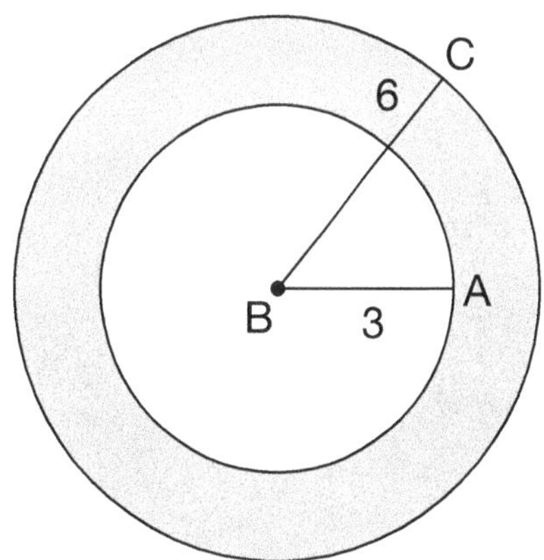

Small Circle Area = πr^2

= 9π

Big Circle Area = πr^2

= 36π

Shaded Part Area = $36\pi - 9\pi$

= 27π

Correct Answer : D

11. Solution:

$(x - h)^2 + (y - k)^2 = r^2$

$(x - 3)^2 + (y + 2)^2 = 5^2$

$r^2 = 25$

$r = 5$

Correct Answer : D

12. Solution:

$$A(ABC) = \frac{8 \cdot 8}{2} = 32$$

Area of Quarter of Circle

$$= \frac{\pi r^2}{4} = \frac{64\pi}{4} = 16\pi$$

Shaded Area $= 16\pi - 32$

Correct Answer : B

13. Solution:

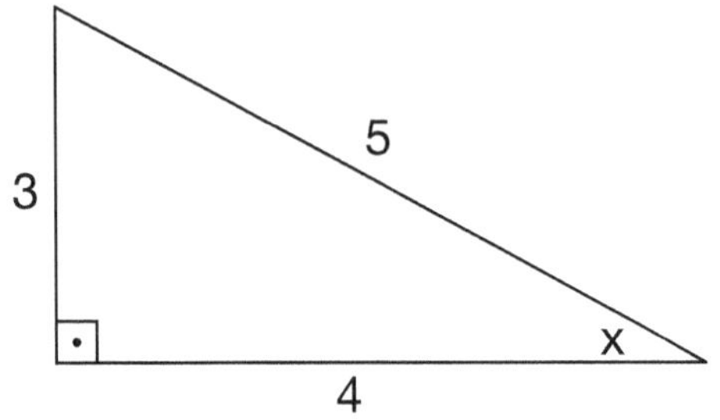

$$\text{Cotx} = \frac{\text{Adj}}{\text{Opposite}}$$

$$\text{Cotx} = \frac{4}{3}$$

Correct Answer : B

14. Solution:

(1, 2) and (–1, –2)

$$\text{Slope} = \frac{y_2 - y_1}{x_2 - x_1} = \frac{-2-2}{-1-1}$$

$$\text{Slope } \frac{-4}{-2} = 2$$

y = mx + b

y = 2x + b, (1, 2)

2 = 2 + b

0 = b

y = 2x

Correct Answer : B

American Math Academy

ANSWER KEYS

SAMPLE TEST ANSWER KEY

1	2	3	4	5	6	7	8	9	10	11	12
D	B	C	D	B	D	D	B	B	B	C	D
13	**14**	**15**	**16**	**17**	**18**	**19**	**20**	**21**	**22**	**23**	**24**
D	C	C	D	A	C	C	C	A	B	D	B
25	**26**	**27**	**28**	**29**	**30**	**31**	**32**	**33**	**34**		
C	C	A	C	B	D	C	A	C	D		

LOCICAL REASONING TEST ANSWER KEY

1	2	3	4	5
C	A	B	D	C

6

- **Converse:** If a bird is not a penguin, then it can sing.
- **Inverse:** If a bird cannot sing, then it is a penguin.
- **Contrapositive:** If a bird is a penguin, then it cannot sing.

7

- **Converse:** If a car does not use gasoline, then it is electric.
- **Inverse:** If a car is not electric, then it uses gasoline.
- **Contrapositive:** If a car uses gasoline, then it is not electric.

8

- **Converse:** If a shape is a square, then it has four equal sides.
- **Inverse:** If a shape does not have four equal sides, then it is not a square.
- **Contrapositive:** If a shape is not a square, then it does not have four equal sides.

9

- **Converse:** If a planet is not outside our solar system, then it orbits the sun.
- **Inverse:** If a planet does not orbit the sun, then it is outside our solar system.
- **Contrapositive:** If a planet is outside our solar system, then it does not orbit the sun.

10

- **Converse:** If a number is not prime, then it is even.
- **Inverse:** If a number is odd, then it is prime.
- **Contrapositive:** If a number is prime, then it is odd.

ANSWER KEYS

ANGLE RELATIONSHIPS TEST ANSWER KEY

1	2	3	4	5	6	7	8	9	10	11	12
D	B	D	C	A	B	C	D	B	A	C	B

INTERIOR AND EXTERIOR ANGLES TEST ANSWER KEY

1	2	3	4	5	6	7	8
$m\angle 1=70^\circ$	$m\angle 3=110^\circ$	$\angle 4\cong\angle 7$, $\angle 2\cong\angle 5$	$\angle 1$, $\angle 3$, $\angle 6$, $\angle 8$	$\angle 1\cong\angle 5$, $\angle 3\cong\angle 7$, $\angle 4\cong\angle 8$, $\angle 2\cong\angle 6$	$\angle 1$ and $\angle 3$, $\angle 2$ and $\angle 4$, $\angle 5$ and $\angle 7$, $\angle 6$ and $\angle 8$	$m\angle 7 = 105^\circ$	$m\angle 4 = 125^\circ$

SEGMENT ADDITION POSTULATE TEST ANSWER KEY

1	2	3	4	5	6
B	B	A	B	D	B

ANGLE BISECTORS TEST ANSWER KEY

1	2	3	4
C	D	A	C

INTERIOR AND EXTERIOR TRIANGLES TEST ANSWER KEY

1	2	3	4	5
C	A	B	D	B

ANSWER KEYS

TRIANGLE INEQUALITIES TEST ANSWER KEY

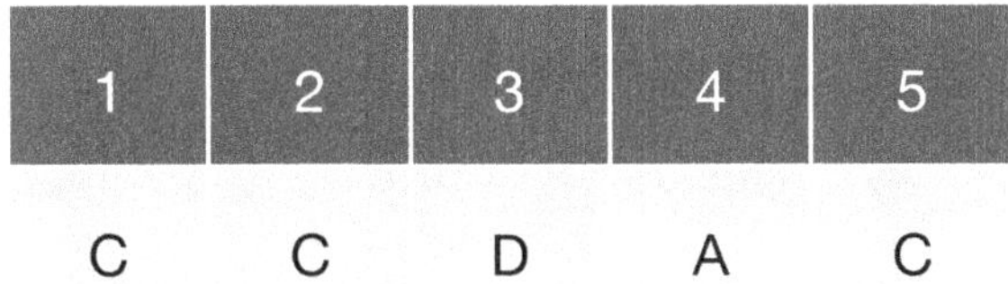

1	2	3	4	5
C	C	D	A	C

SPECIAL RIGHT TRIANGLES TEST ANSWER KEY

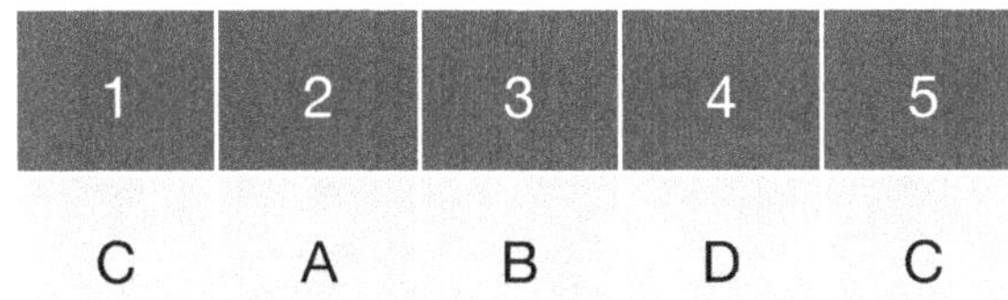

1	2	3	4	5
C	A	B	D	C

PYTHAGOREAN THEOREM TEST ANSWER KEY

1	2	3	4	5
D	D	A	D	D

CLASSIFYING TRIANGLES TEST ANSWER KEY

1	2	3	4	5
B	D	C	A	D

COORDINATE PLANE TEST ANSWER KEY

1	2	3	4	5
B	D	D	D	D

SLOPE AND SLOPE INTERCEPT FORM TEST ANSWER KEY

1	2	3	4	5	6
B	A	B	C	D	C

ANSWER KEYS

SIMILARITY THEOREM TEST ANSWER KEY

1	2	3	4	5	6
A	A	C	C	A	C

AREA AND PERIMETER OF TRIANGLE TEST ANSWER KEY

1	2	3	4	5	6	7
A	B	C	C	D	B	B

AREA AND PERIMETER OF QUADRILATERALS TEST ANSWER KEY

1	2	3	4	5	6	7	8
C	A	D	B	B	C	D	B

MIDPOINT AND DISTANCE TEST ANSWER KEY

1	2	3	4	5
C	B	A	C	A

TRANSFORMATIONS TEST ANSWER KEY

1	2	3	4	5
D	B	D	C	C

ANSWER KEYS

CIRCLES TEST ANSWER KEY

1	2	3	4	5	6	7	8
B	D	A	B	B	A	C	C

TRIGONOMETRY TEST ANSWER KEY

1	2	3	4	5	6	7
D	A	B	B	D	A	B

VOLUME TEST ANSWER KEY

1	2	3	4	5	6
D	C	B	C	B	B

MIXED REVIEW TEST 1 ANSWER KEY

1	2	3	4	5	6	7	8	9	10	11	12
B	B	A	B	B	A	B	C	A	A	A	A
13	14	15	16	17	18	19	20	21	22	23	24
B	D	C	B	D	C	A	A	A	A	A	D

MIXED REVIEW TEST 2 ANSWER KEY

1	2	3	4	5	6	7	8	9	10	11	12
B	A	B	D	D	C	A	C	C	C	B	A
13	14	15	16	17	18	19	20	21	22	23	24
C	A	A	C	C	A	B	B	A	A	A	A

ANSWER KEYS

MIXED REVIEW TEST 3 ANSWER KEY

1	2	3	4	5	6	7	8	9	10	11	12
C	B	B	A	B	C	C	C	A	A	C	A
13	14	15	16	17	18	19	20	21	22	23	24
A	A	C	C	C	C	A	A	B	A	A	A

MIXED REVIEW TEST 4 ANSWER KEY

1	2	3	4	5	6	7	8	9	10	11	12
A	A	D	B	B	A	C	C	B	D	A	A
13	14	15	16	17	18	19	20	21	22	23	
A	B	C	A	D	A	A	A	D	A	B	

FINAL TEST ANSWER KEY

1	2	3	4	5	6	7	8	9	10	11	12	13	14
A	D	C	C	D	A	A	C	A	D	D	B	B	B

NOTE

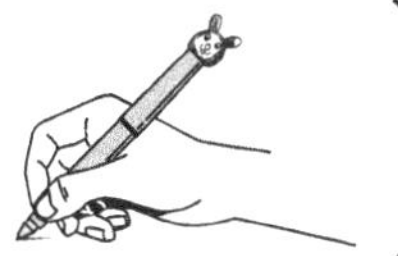

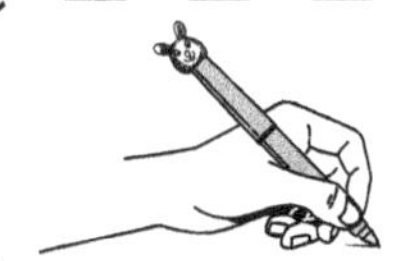

NOTE

NOTE

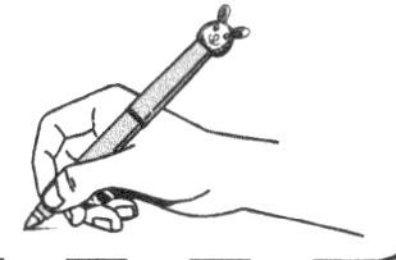

www.ingramcontent.com/pod-product-compliance
Lightning Source LLC
Chambersburg PA
CBHW060601120726
48002CB00010B/2766

9798869116550